SACRED SCIENCE?

SACRED SCIENCE?

On science and its interrelations with religious worldviews

edited by:

Simen Andersen Øyen

Tone Lund-Olsen

Nora Sørensen Vaage

Wageningen Academic
Publishers

ISBN: 978-90-8686-196-5
e-ISBN: 978-90-8686-752-3
DOI: 10.3920/978-90-8686-752-3

First published, 2012

This book is dedicated to Gunnar Skirbekk,
founder of the Centre for the Study of the Sciences and the Humanities,
University of Bergen

Editorial advisory board

Preface

The Centre for the Study of the Sciences and the Humanities (SVT) is celebrating its 25th anniversary in 2012, and this book is part of the anniversary celebrations. It aims to offer topics of interest to most researchers with a connection to the research profile of the SVT. But the book is also intended to reach a broader readership, outside of academia.

The Centre for the Study of the Sciences and the Humanities, University of Bergen, originated in a diaspora from the Department of Philosophy. The Centre was established in 1987 as a *permanent institute*, with standard obligations in research, teaching, popularization and dissemination. Due to the definition of "*vitskapsteori*", the academic profile of the SVT differs from that of "science studies" in the Anglo-American sense: all university disciplines are included as "research objects". The research profile of the SVT is broad, with a focus on critical reflection over the science-society relationship and the ethical and societal aspects of science and technology. The theme of religion and science in modern society is one of the problems that the SVT finds interesting.

All the chapters in this book have been through an anonymous peer review.

First of all, we would like to acknowledge Shijun Tong's participation in this book. He was unfortunately prevented from handing in his chapter, which discussed various aspects of scientific research on Chinese spiritual life. It is a great loss to this book.

This book was made possible by the efforts and devotion of many people. We would like to thank all of our contributors. Many thanks also to our eminent scientific advisory board consisting of Paola de Cuzzani, Jonas Jakobsen, Carl Walter Matthias Kaiser, Torjus Midtgarden, Jørgen Pedersen, Rasmus Slaattelid and Jan Helge Solbakk. Without you this book would not have been possible.

The Centre for the Study of the Sciences and the Humanities is a very congenial environment to work in, and we would like to give particular thanks to Signe Solberg and Line Nævdal, who have been part of this project since the start. Janne Cecilie Johansen's contribution to this book in the area of translation, insightful comments and editing should also be emphasized. We are also grateful to Gordon Lynch, Marianne Bøe, Kjetil Rommetveit and Johannes Servan. Lastly, but most importantly, we would like to thank our publisher Mike Jacobs for the interest and support he has shown throughout the whole of the project.

For any errors or omissions, we accept full responsibility.

Bergen, February 2012
Simen Andersen Øyen, Tone Lund-Olsen and Nora Sørensen Vaage

Table of contents

Chapter 1: Scientific worldviews, religious minds

Some introductory reflections

Simen Andersen Øyen, Nora Sørensen Vaage and Tone Lund-Olsen

Throughout most of the history of science, academic practice has been regarded as a fundamentally rational process. But is it really that rational? When the first universities in Europe were founded in the twelfth century, a ceremony resembling the modern doctoral defence was quickly established. The candidate was presented with two opponents, who examined the strengths and weaknesses of his dissertation. Apart from the removal of a somewhat barbarian precaution – the candidate's oath included a promise not to try and kill the opponents should he fail – the structure of the doctoral defenses has not changed significantly to this day. The ceremonial structure closely resembles that of religious *rites of passage*. The traditionalist nature of the doctoral defence not only illustrates that the university is a conservative institution, but also how it is interrelated with the world of symbols, rituals and ceremonies (Krogh *et al.* 2003). What are the most salient symbols of academia today? This question is too complex to be answered in a few sentences, but there seems to us to be a strong inclination in our time towards what Fisher called the "symbols of achievement" (Fisher in Jegerstedt 2011). Success is primarily measured by the number of publications and prizes, funding and international rankings, which seem to be valued over actual achievement. This illustrates how rituals and symbols remain at the core of modern life and are defining aspects of academia – the institution which is the actual symbol of the rationalization and the secularization of Western societies (Weber). The secular is sacred. All social action is also performative, governed by imaginaries, narratives and collective representations. All that is profane is also holy, to paraphrase Marx.

Science and religion are often viewed as dichotomies, one being associated with systematized, empirical observation, the other more with spirituality, morality and human existence. This can be seen in the thesis of secularism and the presumed rationalization of the modern world. The separation of science and religion has been productive and effective in generating extensive technological and scientific "innovation", but this is an artificial division. A considerable number of religious believers relate to the discoveries of science as confirmation of their present beliefs, and do not find any conflict between the words of God and the teachings of, say, evolution.

Religion used to fill many of the functions that are today claimed by science, of providing answers to the great mysteries of life. Science understood in its broadest terms – as everything between the disciplined study of the natural world to reflecting the narratives and symbolism of complex societies – seeks to achieve a privileged epistemological position over experience based in the life world. Religious beliefs can be seen as some of the most important ways in which human beings have sought reassurance when experiencing that they are not in control of their lives. Religion is a way of meeting the threat of meaninglessness, and has traditionally transmitted wisdom about what is of essential value and ultimate meaning as a guide for human living. People can find the reassurance they are looking for in approaching what they feel is another dimension, not accessible to ordinary, everyday experience. This alternate reality is infinite in scope and is expressed primarily through symbolic expressions in community with others. However, the religions as well as the sciences are continuously changing and evolving to fit the problems arising from the new lifestyles, new technology and new expectations of contemporary individuals.

We might suggest that the *methods* employed by scientists and theologists have often differed greatly, while their *cognitive and social goals* have frequently overlapped. Also, in many cases, both scientists and religious believers are concerned with the same phenomena, but arrive at very differing conclusions. They are separated by the practices, symbols, rituals, institutions and communities which sustain their ways of interpreting and perceiving the world. This is as true for the many disagreeing fractions within each field, as it is for the fields as collective concepts.

Is a system that offers answers to the questions about the fundamental nature of the universe, about the place of human beings in that universe, without answering to any supernatural beings, still a religion? In this context, the notion of a *religious belief* poses the salient question: what turns *belief* into *religious faith*? What makes a person consider an action, an object or a subject as sacred? And can beliefs be considered scientifically? Can science, itself, be considered a religion or a belief?

Science as religion?

Academics have a tendency to believe that the sciences and universities ensure well-documented knowledge, and can function as a democratic-moral corrective to the world outside and that technology and science will solve our fundamental problems. This is an implicit dogma. In certain cases, this dogma approaches the level of religious belief, in which institutionalized science plays the role of the church of the medieval ages (Feyerabend 2011). Science's authority and prestige allows for little in the way of alternate approaches not founded in empirical verification. For instance, holistic medicine and postmodern gender researchers have long been rejected as unscientific. Has truth been colonized by science, or is it only one ideology among others?

In opposition to a view on science and academic practice as a consistent, homogenous, continuously accumulative project of Enlightenment, this practice can be seen as a disciplinarian and politicized one. Gilles Deleuze and Félix Guattari (1980) would say that academic terminology is a kind of sociolect. Knowledge is consequently not just a rule of conduct for political and strategic action but also a basis for legitimacy, rule, dominance, etc. Hence, the extensive knowledge-based Western society – with its historically structural, social and economical inequalities – will always be part of and based on several other discourses which can be described as hegemonic. Consumer-based capitalism is one of these discourses, fragmented though it is. The academic practices are also in this regard an unmanageable quantity. How can standards of rationality and scientifically regulative ideas, for instance in the development of new technologies, be defended today? Have scientific disciplines become so specialized and "operationally closed" that they have constructed insurmountable barriers to other disciplines as well as the general public? If this is the case, can science be controlled and politically guided? It is in this tension, between the belief in Enlightenment and in disciplinarian processes, between the Humboldt ideal and the imperatives of the marked and the state, between sacred and profane, between religion and rationality, that the academic practice as systematic, institutionalized knowledge is constituted.

Science and religious worldviews intertwined

Long-standing notions of the relationships between science and religion have presupposed that the modern industrialized society will relegate religious activity and thinking to only a marginalized position; the public sphere will become increasingly dominated by scientific rationality, while religion will have to struggle for a place within individuals' private spheres (Berger 1967, Durkheim 2008, Weber 2002). However, this secularization thesis was seen by many to have been proved wrong with the widespread religious revival and rise in new religious movements that have occurred throughout the last decades. What does that imply for our understanding of the relationship between science and religion?

One can now observe attempts to reconcile natural-scientific accounts of the world with traditional religious beliefs as well as scientists that employ scientific methodology to argue that religious faith is a delusion (see amongst others Dawkins 2009). Do our concepts of science and religion require revitalization? Have our frameworks for understanding both of them been too narrow, and do we see a naturalization of the religions and a theologization, or politicization, of the sciences? Can we operate with clear dividing lines between religion, when it concerns existential questions, and scientific inquiry in the understanding and explanation of natural phenomena, social processes, etc.? All knowledge, both scientific and religious, should be understood within a cultural and historical context. With the advent of modern science, scientific development claimed to hold the key to real knowledge. Religious worldviews clashed with the ideas of natural science. The production of knowledge became contested terrain, and it still is today. A significant question for our time is: what are the changing conditions for the production of knowledge? The questions posed in this introduction are some of the ones our writers set out to explore, keeping in mind the risk of simplification and the danger of being seen as historically near-sighted or ethnocentric.

Sacred science?

In the chapter "Science and religion? Eight crucial points", philosopher Gunnar Skirbekk addresses the questions whether science can solve our fundamental problems and what the relationship is between science and religion in modern societies. He advocates that both science and religion have to be subject to a continuously self-critical discussion in order to facilitate a constructive act of improvement. Religion stands in a critical interplay with the various sciences. In our pluralistic societies there is a need for informed and enlightened criticism, which implies a critique of dogmatism as well as relativism within the sciences. A critique of the basic validity claims of the religions should be put forth and adjusted to our "risk society", or else religious fundamentalists risk being unintentionally blasphemous by believing in religious dogmas concerning how we eat and dress. Why should God be interested in whether we eat pork or wear silk shirts, when a huge number of people are starving?

Sociologist Cathrine Holst, in "What is epistocracy? Dimensions of knowledge-based rule", sees the role of knowledge in political decision-making through the lenses provided by the concept of "epistocracy" – a "rule of the knowers" – and in comparison and in a conceptual relationship to democracy and technocracy. In the same way, but as an alternative to a "theocracy" or "rule of priests", epistocracy legitimizes the political system. Epistocracy as an analytical concept can define and explain some important aspects of political decision-making of late modern societies and facilitate both descriptive empirical work and normative assessments. But it can also revitalize the role of knowledge in political decision-making as a central topic in the Western history of political theory, and address how it should be applied today.

Philosopher of science Roger Strand, in "Doubt has been eliminated", discusses the relationship between science and life philosophies (*livssyn*). His particular focus is dealings with the issue of climate change in contemporary society. Can one appropriately talk about life philosophies while discussing climate change? Strand takes as his example a speech made by the UN Secretary-General's Special Envoy on Climate Change, Gro Harlem Brundtland, before the UN Commission on Sustainable Development in 2007. In this speech, Brundtland proposed that the time for action has come, claiming support from two recent scientific reports on climate change. According to Strand, Brundtland's argument that Science is Right is an example of what he terms an "unscientific belief in science". Moreover, this belief was utilized for political gain. According to Strand, there can be a scientific belief in Science, but that belief cannot be too dogmatic or too hostile towards criticism without becoming unscientific. He then poses the question: if Science is not the source of authority for this type of belief in Science, what exactly is its source?

A similar critique of a religious belief in science is presented by philosopher Simen A. Øyen. In "The religious belief in rationality, science and democracy", Øyen criticizes what he finds is a tendency to believe that the sciences and universities can be a democratic-moral corrective to the political sphere, and that technology and science will solve our fundamental problems. These problems, he claims, are often created by the sciences themselves. Øyen questions whether institutionalized science functions as a democratic-moral corrective or whether it is in certain ways a theoretical support for a specific political regime – the liberalistic democracy. He offers a critique of the rational-individualistic Enlightenment tradition, and its modified expression in the communicative philosophy of Habermas. These questions are analyzed especially in the context of the theories of Paul Feyerabend. How is the dogma of science and the universities as a democratic-moral corrective equivalent to a religious belief? And how can we criticize this dogma without turning to relativism? Øyen presents a conception of modernity as disciplining processes, suggesting that academic practices, and the ideals springing out of them, should rather be seen as normative regimes or moral imaginaries.

Philosopher and psychologist Ole Jacob Madsen also dissolves the strict distinction between science and religion. In the chapter "Psychology as science or psychology as religion: Historical assumptions and consequences for the present" he poses the question of whether scientific knowledge, in the case of psychology, contains religious assumptions and shares some important features with religion. He asks if psychology has replaced, especially, Protestant Christianity, in providing a view on self and society, for instance by offering new rationales for human suffering, which is and has been among the most vital aspects of all world religions. By examining the theoretical and ethical foundations of psychology both as a natural science with historical roots to nineteenth-century Germany, and as a gradual continuation of the individualistic roots dating back to the sixteenth century, Madsen sees these tensions between the scientific and religious in the context of a late modern, globalized therapeutic culture. Whether we choose to consider psychology as science or as religion affects how we reflect on psychology and its range of influence.

In the chapter "Science without God", physicist and philosopher Ragnar Fjelland points to the abundance of literature that uses science either for or against religion. His main focus is on a group of biologists that gives an evolutionary account of religion and that argues that religion is incompatible with science. Fjelland categorizes these biologists and their arguments as representatives of what he calls scientific fundamentalism and claims they have a lot in common with religious fundamentalists in their insistence on the natural sciences as providers of the "one true perspective". Fjelland, on the other hand, argues that religious questions are simply outside the scope of science.

Literary critic and theorist Barbara Herrnstein Smith, on the other hand, takes a different view on where evolution will lead religion. In the chapter "Science and religion, natural and unnatural" she problematizes the recent years' wave of cognitive-evolutionary accounts of the origins of religion. These accounts explain religious concepts as products of the unconscious operations of innate, universal mental mechanisms that evolved in humans under Stone Age conditions. Religious concepts are, according to this strand of research, inherently attractive to the human mind and therefore considered cognitively natural whereas scientific thought is seen as cognitively "unnatural". Smith questions the sharp contrast between science and religion set up by the spokesmen of the cognitive-evolutionary approaches, and alerts us to similarities in the cognitive springs of science and religion that are elsewhere seen largely in terms of difference and opposition.

How do recent technological advances affect our ideas of what science and religion can and should be? In the chapter "Immortality: an essay on science, technology and religion", philosopher of science Kjetil Rommetveit addresses attempts in recent years to engineer immortality and the transhumanist movement. According to Rommetveit, prolonged life has become a serious goal for research and innovation to the extent that it is becoming a *leitmotif* for the 21st century. He takes the reader through a number of epochs and some of their ways of imagining immortality, from the ancient Greeks to the present day's proponents of transhumanism, to put to the test the hypothesis that immortality, whilst clearly an esoteric and transcendent phenomenon, also possesses a number of earthly, social and cultural characteristics.

Bioethicist and educator Michael J. Reiss, in "What should be the role of religion in science education and bioethics?" takes two quite different stands as to the roles of religion in science education and bioethics, respectively. In addressing science education, he concentrates on the issue of whether Creationism should be discussed in the science class. He argues that a discussion of Creationism does not imply a legitimization of the idea, and points out that a number of science students can be supposed to hold Creationist views. This, he suggests, is a good reason to present students with the scientific consensus about evolution. Considering bioethics, Reiss argues that ethical questions cannot be positively decided by reason alone. In a multicultural society, well-argued viewpoints should always be heard, be they religious, or not.

In his Current Commentary, entitled "The arc of civil liberation", sociologist Jeffrey C. Alexander offers some perspectives on our present time, taking as his examples Barack Obama, Tahrir Square, and the Occupy Wall Street movement. Alexander argues that, even though our modern society is often perceived as rational, we still need broad, metaphysical beliefs that are not proven empirically. Many of the so-called secular movements of our time embody a powerful symbolic message, as the three examples, in their different ways, illustrate. The symbolic eruptions of utopian possibility form a narrative arc of civil liberation. The potent icons and symbols created by these movements show us their sacred nature.

In this book, our writers set out to investigate whether the symbols of academia may in some cases take on a quality of sacrality, whether "priests of knowledge" may legitimately hold political or moral sway in society (the rule of knowers and experts), whether religion has a place in scientific contexts, and a selection of other questions concerning science and its relations to religious belief. We hope and believe that the following pages will provide some perspectives on these complex aspects of the human social and cultural life world.

Bibliography

Berger, P. 1967. *The Sacred Canopy: Elements of a Sociological Theory of Religion.* New York: Doubleday.

Dawkins, R. 2006. *The God Delusion.* London: Bantam Press.

Deleuze, G. & F. Guattari. 1983. *Anti-Oedipus: Capitalism and Schizophrenia.* Minneapolis, Minn.: University of Minnesota Press.

Durkheim, É. 2008. *The elementary forms of religious life.* Oxford: Oxford University Press.

Feyerabend, P. 2011. *The Tyranny of Science.* Cambridge: Polity Press.

Fisher, M. 2009. *Capitalist Realism. Is There No Alternative?* Winchester, UK and Washington DC: Zero Books.

Jegerstedt, K. 2011. "Teaching in the End Times": Humaniora og kunnskapsproduksjon i globaliseringens tidsalder. In S. Øyen *et al.* (eds.). *Humanioras fremtid.* Oslo: Cappelen Akademiske Forlag. 113-137.

Krogh, T. *et al.* 2003. *Historie, forståelse og fortolkning. De historisk-filosofiske fags fremvekst og arbeidsmåter.* Oslo: Gyldendal Norsk Forlag.

Weber, M. 2002. *The Protestant Ethic and the Spirit of Capitalism.* Los Angeles: Blackwell Publishing.

Chapter 2: Science and religion?

Eight crucial points

Gunnar Skirbekk

The editors asked us to reflect on questions such as "Can science solve our fundamental problems? What is the relationship between science and religion in modern societies?" Big questions, indeed. In this chapter, I shall give my answer by referring to four points concerning the sciences and four points about science and religion.

Common to all sciences: informed and self-critical argumentation

At the outset, we need to explain what we mean by these terms: here I understand "science" in the sense of "*Wissenschaft*", that is, not merely as natural science (*Naturwissenschaft*), but, broadly speaking, as all the disciplines at a full-scale university. As a reminder I shall often use the term in plural, not "science", but the "sciences". I shall return to the term "religion" below, but first some remarks on the sciences.

When talking about the sciences "solving our problems" it is often assumed that the sciences are useful in an instrumental sense, that is, as a tool for improved governance and economic growth. However, not all sciences are instrumentally useful in a Hempelian sense (Hempel 1949). In general, the humanities are not (Skirbekk 2007). Their potential value, and use, is of another kind.

Hence, "value production" in terms of instrumental usefulness for politics or economy cannot be the common legitimation or the common denominator of all the various sciences. They are different both as to the status of their research results and as to what the researchers are doing, for instance in labs, fieldwork or libraries. Yet, in all disciplines, however different they may be from one another, researchers *in spe* have to defend their theses in a doctoral disputation. In this sense, not instrumental usefulness, but argumentative and self-critical discussion is that which all university disciplines have in common.

Surely, in reality there are great differences as to the extent and depth of the argumentative activity, not least as to its self-critical and reflexive part. Moreover, in scientific communities there is also a demand for originality, for new conceptions, methods and techniques, and there are elements of discretion and of tacit competences. Nevertheless, within all of this, at crucial moments a competent and professional "give and take" of reasons is required.

This is my first point: that which is common to all sciences, at least ideally and potentially, is not instrumental usefulness, but informed and self-critical argumentation.

The need for critical studies of the sciences

Also in a broader scope, beyond the inherent activities of the various disciplines, self-critical discussions are needed; this is true both for the interface between the various disciplines and for the relationship between the sciences and society. These are a few reminders: each discipline has its proper conceptions and perspectives, such as in economics, ecology, sociology, political science, and psychology. Thus, each discipline (and subdiscipline) reveals some phenomena and disregards other phenomena. Moreover, there is no "God's eye view", a kind of meta-science and meta-language that encompasses all the different conceptions and disciplinary perspectives in a higher semantic synthesis. As finite human beings all we have is going-between and reflecting-upon the various disciplines, activities that require some "double competence", some insight into what is going on in the disciplines under consideration.

Now, because of the perspectivist nature of the various sciences we may consider two major challenges in the relationship between the sciences and society, not least for political actions: firstly, there is always a danger that one discipline (or even subdiscipline) may acquire a predominant position (among political agents, or in the public awareness), at the sacrifice of other disciplines that reveal other aspects of the problems we are facing. A flagrant case is the discrepancy between the dominant position of (short-term) neoliberal economics and the weaker position of (long-term) ecology. This is a general challenge in modern societies with extensive disciplinary differentiations (specializations). For instance, we may talk about "economism" in cases when some economic disciplines get "the upper hand" at the sacrifice of other relevant disciplines, and about "biologism" in similar cases for biological (or neuroscientific) disciplines, or about "contextualism" in cases when contextualist cultural studies get a dominant position, disregarding the deeper epistemic claims, including their own. Secondly, due to the perspectivist nature of the various sciences there is also a danger that the practitioners in one field (say, nuclear physics or biochemistry) do not envisage the potentially unintended consequences of their own research, a challenge that requires other disciplinary perspectives, e.g. from the social sciences.

This means that there is a "struggle between the faculties" (the disciplines) – not merely within multidisciplinary institutions such as universities, but also in society at large, a struggle that is often related to strong political and economic interests. Hence, we should always keep a critical eye on the possible power-relatedness of the various scientific disciplines and on how they are used and implemented in modern societies. Added to this, in modern societies we have "Big Science", based on strong economic agents and institutions in politics and industry, the military industry included.

In addition to the extensive differentiation of disciplines and specialties, there is a massive growth in the number of researchers and research institutions, including published reports and results. Thus, it has become increasingly difficult to obtain a professional overview of what is going on in a traditional discipline, and it has become increasingly tempting to remain within one's own narrowly conceived professional network. As a consequence one avoids disturbing and critical remarks from researchers in neighboring fields and from enlightened and interested laypeople. In this sense, we have a new kind of deficient overview and lack of transparency (*eine neue Unübersichtlichkeit*, cf. Habermas 1985).

All in all, this means that there is a need for critical and informed discussions of the various sciences and of their use and misuse in modern societies – in short, a need for critical and informed "theory of the sciences" (*vitskapsteori*, cf. NAVF 1976), by someone who has a reasonable knowledge of the disciplines under consideration. Since all the sciences should be considered, not only the natural sciences, these "studies of the sciences" will also include a self-reflective and self-critical activity as to their own validity claims; in this sense they have to include an "internal" perspective, in an interplay with the "external" ones. This is my second point: due to the scientific plurality and its interwovenness with various activities and agents in modern societies there is an inherent and urgent need for critical studies of the sciences, for a self-critical critique of the sciences (*Wissenschaftskritik*).

A need for improvement

At the outset Karl Popper (1963) had a clear-cut demarcation for what defined a science: falsifiability. On the other hand, for postmodernists the sciences are seen as social activities not to be neatly distinguished from other social activities. Counter to this, Merton (1949) had his ideal-type norms for scientific research (the CUDOS, the scientific ethos; rephrased as communalism, universality, disinterestedness, and organized skepticism).

Above we have indicated that informed and self-critical discussion should be seen as a common denominator for the sciences broadly defined. However, this is not a demarcation line between scientific research (in the broad sense) on the one hand and other social activities on the other. In modern science-based and technology-based societies there are a huge number of science-based professions and activities outside the realm of scientific research, from engineering to teaching and also innumerable activities in our daily life. Moreover, in modern democratic societies there is a need for enlightened discussions in public space, as a precondition for reasonable and fair political decisions. Furthermore, in modern pluralistic societies, where fewer activities are predetermined by a given tradition or religion, the only way, for finite and fallible human beings, to be reasonably sure that one's own opinions are liable, is open and honest discussion with other people (cf. Mill 2010).

In other words, science-based and science-related activities are spread out into society at large, beyond the realm of scientific research; moreover, the need for enlightened argumentation and discussion permeates modern democratic societies, beyond the realm of scientific research. In this sense, the idea of a demarcation "line" would be distracting, and hence we should rather talk in terms of gradual transitions and alternative versions – in short, we should talk gradualistically, not merely dichotomically.

However, this does not mean that an ideal-type distinction between scientific and other social activities is simply obsolete and "deconstructed" (in postmodernist terminology). To make the point we may compare this distinction with the ideal-type distinction between health and disease. Surely, it makes sense to talk in gradual terms; we are for the most part more or less healthy or more or less diseased in different ways. But by talking in gradual terms we somehow presuppose an analytical distinction between health and disease. The underlying point about what is ideal and what is undesirable could often be taken a step further. In most cases, the practical concern is that of healing an injury or of avoiding a disease, not that of perfection (with the exception of affluent capitalist societies in search of increased consumption beyond the level of basic needs). In short, the aim is in most cases that of improving the situation, getting away from that which is seen as negative. In this respect, we could indicate a similarity with scientific activities, especially in cases of enlightened and open discussion among fallible and reasonable human beings: in these cases, the main point is that of trying to improve one's opinions and basis for action, not to reach the final Truth.

This is my third point: the ethos of enlightened and self-critical discussion does not represent a demarcation line for scientific research, nor for scientific activities in a broader sense. In modern societies, this ethos indicates a general need for improvement, away from that which is conceived as less reasonable toward that which seems to be better, as a communicative and gradual search for better reasons.

Science: part of the problem, part of the solution

When the sciences are defined as above, to what extent could they be said to "solve our fundamental problems"? First, the sciences themselves, conceived as scientific research isolated from societal agents and institutions, can hardly solve any practical problems, only theoretical ones. On the other hand, theoretical knowledge can certainly have an impact on our self-knowledge and our opinions about the world, by its own force, as it were, without any external plans or projects; this is true for insight stemming from heliocentric astronomy and Darwinian theory of the origin of species, and from historical interpretations of religious scripture and Freudian theories of sexuality, just to mention a few.

However, even though scientific research and results by themselves cannot solve our practical problems – this holds true for most of the urgent problems related to economy and ecology and to governance and welfare politics – scientific research can contribute to the solution of practical problems when it is adequately internalized or implemented by suitable agents and institutions. The question as to how this could best be done is a matter of practical experience and discretion, often to be combined with various kinds of scientific insight, especially from the social sciences.

Urgent problems, such as those related to renewable energy and climate change, to the future supply of fresh water and food, and to unsustainable consumption and reproduction, are already utterly complex at the epistemic level, for instance in the sense that various disciplines are required. How do we decide what kind of discipline and knowledge is required in the various cases? How do we decide whether there is an unreasonable dominance by some disciplines and their disciplinary perspectives at the expense of other disciplines that might also contribute to a better understanding of the problem under consideration? And again, what about unintended consequences (often unimagined in one's own perspective)? What about epistemic uncertainty? And what about the danger of various kinds of power-related influence on research processes and research reports? These are problems already at the epistemic level. Then we have the complexity and challenges at the institutional level, including the danger related to pressure groups and special economic interests and political agents, including military and religious organizations and agents. Surely, for this reason there is a permanent need for a critical and self-critical awareness of epistemic challenges as well as of institutional shortcomings and irregular power-relations.

On the other hand, it won't work without institutions and agents. But then there is a decisive difference between irregular power-relations without fair and reasonable legitimation, and power-relations that are regular and regulated, for instance by institutional division of power and the rule of law, and that are thus to be seen as legitimate power-relations.

Constitutional democracy and democratic law-making are seen as legitimate institutions, fostering legitimate decisions. The paradigm case is a self-contained society where those who make the laws are those for whom the laws apply and for nobody else, and where those who give the laws understand what they are doing, including the implications and long-term consequences of what they have decided. This is the principle of popular sovereignty. Laws and other major decisions are legitimate when they have been agreed upon by all those concerned. This is a bottom-up, not top-down legitimation, be it by a sovereign king or by divine command (the latter becomes problematic when there is more than one confession, as we shall see below).

However, in this paradigm case for democratic decisions and law-making there is an inherent problem when faced with persistent minority constellation. This problem cannot be solved by democratic majority vote. At this point, there is a need for an egalitarian political culture characterized by moderate socio-economic differences and a basic solidarity and mutual trust. Moreover, without a basic trust in procedures and persons, those who lose an election could be reluctant to leave office. Hence, trust is crucial for democratic rule; but trust is something that has to be experienced and internalized by those concerned; it cannot be brought in from the outside, nor can it be installed merely by a decision.

However, in modern democratic societies this paradigm case for the legitimation of democratic decisions and democratic legislation has become more or less obsolete, for three reasons:

1. Space. In a modern globalized world decisions made in one country tend to have consequences and implications for citizens in other countries. Take, for instance, the decisions made in the US and their implications for other countries and their citizens.
2. Time. Due to modern science and technology, and modern institutions, quite a few of the decisions made by our generation have extensive implications and consequences for future generations.
3. Insight. In modern society, based on sciences and technology, most citizens have an insufficient insight into the consequences and implications, for future generations, of the various projects and arrangements that have been

introduced by agents and institutions in our generation. In short, for these three reasons, there is a major discrepancy between the paradigm case of legitimate democratic decisions and many of the decisions that we are making.

What can be done? Just to put a label on the dilemmas:

1. The first dilemma, that of space, is primarily an institutional challenge, which soon becomes a political and normative issue. What kind of political borders are feasible and also desirable?
2. Also, the second dilemma, that of time, is at the outset an institutional challenge, which soon becomes a political and normative issue. Our Western democracies, based on frequent elections, have the great advantage that an unpopular government can be rejected by the voting majority; but it works from a short-term perspective, without an institutional safeguard for the voices of future generations. The same is true of Western capitalism, with a short-term perspective for the economic profit of invested capital. In fact, from a geopolitical perspective, only China seems to have political institutions with a basic concern for the assumed needs of future generations, such as food and renewable energy – not for idealistic reasons, but out of self-interest, since the Chinese regime seems to presume that it shall remain in power for another 20 or 50 years, and it therefore needs to take action now in order to counteract social unrest in the future.
3. The third dilemma, that of adequate insight, is both an institutional and an epistemic challenge. For one thing, it is worthwhile recalling that modern democratic societies have extended and mandatory education for all citizens (in our countries, formally for ten years). Historically, in our country, the legitimation of a mandatory and common education for all citizens was a political move designed to foster an egalitarian political culture, suitable for a parliamentary democracy. But today, in modern societies, this challenge has become even more acute. A short story may illustrate the point:
   It is said that at the time of the attack on the Twin Towers in New York on 9/11 a journalist heard the following utterances by two US citizens watching the whole thing: "This is like Pearl Harbor", the first said. "What's that?" the second person asked. "That was when the Vietnamese attacked us and the Vietnam War started", was the reply. The point of the story is this: these were citizens with the right and responsibility of voting for the US president, a mighty agent with the power to make decisions with deep and long-term implications for many people, at home and abroad. Hence, as a citizen of a democratic society – with the right to vote and speak out, with the right to organize and demonstrate – one does have some joint responsibility, certainly always according to one's own position and capabilities, and certainly only

a minor part, but still, as a citizen of a democratic society one does have some joint responsibility for what is going to happen. This point has a crucial implication: an unnecessary lack of insight into the major challenges of our time is regrettable, both for the individual (all depending on personal resources and positions) and for society at large. The latter means that political and social agents have an obligation to further a good common education and foster an enlightened public space. The former means that each citizen (again according to personal resources) has an obligation to be reasonably updated on major issues. The liberal ideal of a total individual freedom "for anything legal" (as stated in "personal ads"), is outdated in modern democratic risk-societies with some degree of shared responsibility. This means, bluntly stated, that each citizen in a modern democratic society has a basic obligation to improve his own status as an enlightened and autonomous person (in the Kantian sense of *Mündigkeit*). A concern for various kinds of scientific insight, and for a self-critical and argumentative approach, is thus included.

Now, back to the initial question: are the sciences supposed to "solve our fundamental problems"? As a response, I restrict myself to two short remarks:

1. Some of the main problems in modern societies are themselves jointly determined by the sciences and science-based projects and technologies and science-related institutions – though often, it has to be said, by one-sided usage and implementation, for instance to the extent that scientifically one-sided technological and economical projects are given the upper hand, politically and institutionally. Therefore, various sciences are parts of the problem.
2. But since there is obviously no way back to a pre-modern pre-scientific world, we are at the same time obliged to look at the various sciences for reasonable contributions that could be helpful in coping with the main problems in modern societies. This goes not only for our understanding of the actual situation and for our political discretion, but also for institutional and technological arrangements that seem to be beneficial for some of our main problems. Therefore, a critical approach to various sciences represents a part of a reasonable response (not to use the ambitious term "solution").

Two more remarks are required here, recalling some major challenges for university research and education in this respect. In modern mass universities the uniform institutional structure and market-related financial foundation (e.g. the Bologna reforms) counteract both the internal and the interdisciplinary "criticism of the sciences" (*Wissenschaftskritik*), at the same time as many research projects, partly due to these institutional and financial structures, tend to be rather trivial (as in many empirical disciplines) or to disregard the self-referential epistemic challenges in their own projects (as in many interpretive disciplines, thus fostering what in technical terms is called "bullshit", cf. Harry Frankfurt (2005)).

So this is my fourth point: for various reasons the sciences themselves cannot "solve our fundamental problems". To some extent, they are part of the problem. However, when viewed critically and self-critically, in awareness of their inherent differences and challenges, and also in awareness of their common ethos in terms of informed and open discussion seeking better arguments and views, the sciences could definitely be seen as important elements in our complex and fallible dealings with the various urgent problems of our time. Then, what about religion and science, in modern societies? Researchers in the descriptive field of "religion studies" (*religionsvitskap*), different from theology and from the philosophy of religion, view the term "religion" in different ways, as rituals and institutions, as tradition and culture, as belief systems and attitudes. However, we also recall that theology and philosophy are "sciences" in the sense of *Wissenschaften*, that is, as interpretive and argumentative activities, with epistemic validity-claims. We may start with a general observation.

Plurality of religions: a need for clarifying definitions and convincing justifications

In our time there is a pluralism of religions, such as different and often opposing versions of each of the three monotheistic religions, Judaism, Christianity, and Islam, or New Age, Satanism, and witchcraft, old and new, and also other world religions, such as Taoism, Hinduism, and Buddhism, and different forms of religious practices with or without a belief in God or theological theses. For instance, one God, or many, or none? Is God radically separated from the world and humankind, or are there transitions between God and human beings, and between God and the world? Is God benevolent, or evil, or both? Given this pluralism, when we talk about religion, who then has the right to decide, for others, what is included and what is excluded?

This is a semantic point with extensive practical implications, both legally and politically. This open-ended, indeterminate pluralism implies that an appeal for general religious rights (of a legal or economic nature) no longer has a clear and definite meaning. This also holds true for what is said about "religion" in legal texts, such as the UN declaration of human rights. Due to this indeterminate pluralism of "religion", the term has to be defined, and if religion is said to deserve respect and legal rights, that has to be justified in each case, with convincing arguments. In other words, if there are special reasons why a certain religion deserves special respect and support, this has to be shown in each case by arguments that are universally understandable and convincing, that is, by universally valid arguments. In short, due to this semantic pluralism, the reference to something as "religion" is in itself no reason for special respect or concern.

This is my first point: in our societies, there is a plurality of religions, of very different kinds. Hence, there is no reason for respect or support simply because something is taken to be a "religion". To deserve special respect and legal support there has to be a clarifying definition and a commonly convincing justification in favor of that special kind of religion.

In the new age: a close relationship between monotheism and science

"Religion and science" – it goes without saying that it all depends on how the terms are conceived. We have already commented on the term "science", and in the paragraph above we have pointed at the pluralism of "religions". Now, to get started let us focus on some main points in the interplay between science and religion in Western history:

1. During the medieval ages in Western Europe there was a close relationship between theology and philosophy (in many ways the main sciences at the time), and there was certainly an intimate relationship between theology and religion, be it Jewish, Christian, or Islamic. In Platonic (Neoplatonic) and Aristotelian philosophy, there were major theological elements, both in ontology and epistemology and in moral and political theory.
2. From late medieval ages into the new age, up to the eighteenth century, there was similarly a close relationship between monotheistic theology and religion on the one hand and the emerging new natural sciences on the other (cf. Shapin 1996). In this connection there were two underlying images:
   a. The narrative of the Two Books: there were the Holy Scriptures, written by God and to be interpreted by the theologians, and the Book of Nature, written by God in mathematical symbols, to be discovered and reformulated, in a mathematical language, by the natural scientists.
   b. The narrative of God as Mechanical Mastermind: the universe, as gigantic mechanical clockwork, has God as its mechanical mastermind, and by their experimental work, it is up to the natural scientists to discover the underlying laws of nature and formulate them in a mathematical language.

This certainly went against some of the main ideas in the Aristotelian philosophy of nature, and thus it went counter to those theologians who insisted on its ontological primacy. The trial against Galileo in 1633 is the paradigmatic case of this controversy. However, generally speaking, the new natural scientists in Western Europe worked against the background of religious images. Atheism was largely a French invention by the end of the eighteenth century! Even a critical enlightenment philosopher like Voltaire was a deist.

My second point on religion and sciences is as follows: at the outset, in the new age, there was a close and positive relationship between religion and science, not least among the new natural scientists.

The inherent need for a critique of religion

Above, we referred to the need for a self-critical critique of the sciences (*Wissenschaftskritik*). Theology and philosophy are included here. However, there is also a need for an informed and self-critical critique of religion (as we already indicated above). But there are different (partly overlapping) kinds of critique of religion, in short:

- Moral-based criticism of religion tries to show that certain forms of religion are morally problematic or rejectable. The targets are utterances and demands found in religious scriptures or doctrines, but also acts and attitudes ascribed to religious persons and institutions.
- External criticism of religion tries to show that certain forms of religion are merely epiphenomena, expressions of underlying psychological and social circumstances.
- Internal criticism of religion tries to take literally some of the utterances in certain forms of religion and thereby raise an intellectual criticism of the level of precision and of the truth-claims in that which is said or presupposed.

The role of the various sciences, for the critique of religion, could be summarized in this way:

1. External, causal explanation: referring to social causes, such as in Marx – religion as opium for the people, and thus as false consciousness – or psychological and psychiatric causes, as in Freud – religion as psychological displacement. But then there are also sociological arguments in favor of religion, as in Durkheim, seeing religion as important for social cohesion.
2. Historical positivism: the secularization thesis, e.g. in Comte – who saw history as a development in different stages: the religious, metaphysical, and scientific stages. However, then there are counterarguments, e.g. in late Habermas, who rejects the secularization thesis and emphasizes religion as a resource of insight and values needed in modern societies (Habermas 2005).
3. Natural scientific knowledge: the origin of the universe (paleontology), the structure and scope of the universe (Kepler, Galileo, modern astronomy),

natural scientific explanations (Newton, God as "watchmaker", without miracles and magic), the origin of species (Darwin) on the background of dramatic cosmological events (geo-history). Each one of these challenges demands a theological response. However, when taken together these cases of natural scientific knowledge strongly indicate that the universe is no safe and friendly place.

4. Logical positivism: the theory of knowledge, focusing on the question as to which utterances can be seen as cognitively meaningful, and giving the following answer: only well-formed, empirically verifiable propositions are cognitively meaningful (Alfred Ayer and the Vienna School).

In short, in talking about the relationship between religion and science, these well-known critical arguments should be mentioned. However, these are the kinds of arguments, well-known in modern societies that have been seriously considered by contemporary theologians and for this reason modern university-based theology has been intellectually modernized. Here again there were long learning processes. We may briefly recall some major points.

The new natural sciences, experimental and mathematically formulated as in Newtonian mechanics, were gradually interrelated with technological developments and thereby related to economy, and also to the State, for instance in the development of infrastructure and military technology. Via the structure of their causally explaining methods, these sciences delivered explanations, predictions, and technical maxims (Hempel 1949). Hence, through these new sciences we could obtain better control of natural events. However, in the same period, with new States and a new religious pluralism resulting from divisions between Catholics and Protestant denominations of various kinds, there was also a renewed concern for interpreting disciplines: the interpretation of legal texts in jurisprudence and the interpretation of religious texts in theology. For a text does not interpret itself; it has to be interpreted by somebody. Moreover, there are often different interpretations of the same text. Hence, we are faced with the question: why is my interpretation better than the other interpretations? For a serious answer to this question, one has to give reasons as to why one interpretation is more reliable than another. Moreover, different religious have different Holy Scriptures, and hence we are faced with the question: why are my texts the right ones, and not those of the others? In short, there is an inherent need, within the religions based on Holy Scriptures, to move from interpretation toward rational argumentation. This reminds us of Enlightenment, as in Kant's (1962) famous definition: *sapere aude!* Have the courage to use your own reason, in a self-critical discussion with other people! Moreover, in Kant, the term "critique" does not mean a negative denial (cf. his "critique" of pure and practical reason), but a serious test. Hence, modern societies are science-based, not only by the new natural sciences, but also by renewed interpretive disciplines and self-critical argumentation. However, this has not been recognized by everybody. For instance, Sayyid Qutb (cf. his book *Milestones* (1998)) was in favor of natural sciences, and certainly of his own religious convictions, but he disliked humanities and social sciences. The same is true of people like Ahmadinejad, engineer and religious fundamentalist, and also of influential groups in the United States who conceive of freedom of religion as a freedom from criticism, not as a freedom to criticize. These people embrace the modern sciences of Galileo and Newton, but not the Enlightenment tradition of Voltaire and Kant.

This is now my third point: for the three monotheistic religions, there is an inherent urge for a critique of religion. However, due to the critical interplay between the various sciences and religion, modern university-based theology in the Western world has largely been intellectually modernized.

Modernization of consciousness

But is not religion (whatever it means) beyond the scope of rationality, either because it is deeply personal or because it can only be understood in an internal perspective, that is, by the believers themselves, or maybe by their spokespersons, such as rabbis, priests or imams? There is something to be said for such objections. On the other hand, when it comes to the three monotheistic religions – Judaism, Christianity, and Islam – they do raise universal validity-claims, each one of them, about their Holy Scriptures and about the one and only God. Structurally, on these decisive points, the three monotheistic religions are faced with the same kind of challenges; and consequently, due to these universal validity-claims, they are inherently open to enlightenment and rational criticism:

1. Based on Holy Scriptures: in a modern pluralistic society, we are faced with the fact that there are other interpretations of "my" Holy Scriptures. Hence the question: why are my interpretations the right ones? And we have to realize that there are other people who have other Holy Scriptures. Hence the question: why are my texts the right ones? To answer these questions, reflexivity and reasoning are needed. Self-critical interpretations and reasonable argumentations are required.
2. Monotheism, belief in one God (Jahvé, Allah): for all three monotheistic religions there is only one God (mono-theism), who is at the same time the creator (and supporter) of the world, the lawgiver, the judge, and the executioner. Given that God is almighty, benevolent, and omniscient – he is all good, he knows everything, and he can do whatever he likes – then, when

> faced with major tragedies and disasters (such as the earthquake in Lisbon in 1755) we are faced head on with a major problem: how could God allow this to happen? In theological terms, how can we cope with "the problem of evil"? On this major problem in the three monotheistic religions, there are ongoing discussions, from the Book of Hiob in the Old Testament up to Leibniz and his theodicy in the eighteenth century, and further on in our time. How should we theologically understand and explain major disasters – earthquakes, tsunamis, sudden ice ages or exploding calderas? Where was the voice of Jahvé in Auschwitz? Arguments of free will, or of unavoidable interdependence of good and evil, do not cope satisfactorily with such major disasters and tragedies that do not appear to be necessary (either empirically or logically), and that cannot be understood as divine punishment for sinful acts committed by the victims.

Moreover, at this point there is even a paradoxical danger of "involuntary blasphemy", among uneducated and fundamentalist believers, who regard themselves as true defenders of the right faith: when God (Jahvé, Allah) is conceived of by these people as the sovereign creator who has given us strict laws and rules of behavior, and who at the same time operates as a severe judge and executioner, sending condemned sinners to hell for eternity, then the "problem of evil" reappears as a question as to whether God, conceived of as such a brutal master, in reality acts like a Satan. Thus, their conception of mono-theism looks like a mono-Satanism – and that must surely be seen as blasphemy, even if it is not recognized as such by those who think in these terms. In short, in these cases we have involuntary blasphemy, but blasphemy nevertheless. Moreover, the same holds true for uneducated believers who think that in our dangerous and precarious world, where a huge number of people are starving and are without shelter, the supposedly merciful God is seriously and predominantly interested in what we eat and how we dress – no milk and meat, no pork, and no silk shirt for men! – which in reality means that God has no sense of proportion and thus He appears as a ridiculous figure – a blasphemous view, again a case of involuntary blasphemy.

In short, the three monotheistic religions are similar on these two essential levels:

1. interpretations of sacred texts, and
2. the belief in one God/Jahvé/Allah as creator, legislator, and judge, and hence, they are faced with similar challenges, such as the problem of evil.

Then there are differences between (and within) these religions. For instance, they are dissimilar due to different historical conditions, e.g. as to whether they operated inside or outside the realm of political and military power, or as to how they were interrelated to the institutional and epistemic developments that were parts of early modernization processes. But all three are today faced with the same basic needs for epistemic and institutional adaptation to the positive and necessary demands for an enlightened modernity, in short, for a "modernization of consciousness" (Habermas 2005):

1. A recognition of various kinds of insight and knowledge that are established by sciences and scholarly disciplines, though critically conceived, but are still the best we have. Religious teaching and practices should be adapted accordingly.
2. A self-critical reflection on, and recognition of, the plurality of religions and other "comprehensive doctrines" (Rawls 1993). Religious teaching and practices should be adapted accordingly.
3. An institutional differentiation between the legal system and religion. Reasons given for common coercive laws should be universally understandable and the procedures should be universally acceptable.

In principle these three points are demands for everybody, though in each case dependent on personal background and resources. They are, first and foremost, general demands for political and religious leaders. However, when these demands for a "modernization of consciousness" are not dealt with appropriately, then we do not live up to the main preconditions for modern societies, with their variety of sciences and scholarly disciplines and their institutional differentiations, and with their pluralism.

Above we focused on the need for a self-critical critique of the sciences and now we have focused on a need for a self-critical critique of religion, in its interplay with the sciences in modern societies. It is worth recalling that critique in this connection does not mean rejection. The term should be taken in its Kantian sense: critique as a purifying process, as in the Kantian "critique of pure reason" and "critique of practical reason". The point of the critique in this sense is not a negative act of rejection, but a constructive act of improvement.

As my fourth point I would therefore rearticulate my main message on science and religion: "religion" is no longer a precise term and it is no longer self-evident that "religion" should be respected and legally supported. A normative justification is needed in each case. In short, religion is part of the modern world, though not without a critical interplay with the various sciences, which implies both a self-critical critique of the sciences and a self-critical and purifying critique religion, the latter characterized by a "modernization of consciousness". In this sense, the sciences can at least help us to deal with problems concerning scientific self-understanding and our understanding of religion in a modern world. These are my concluding remarks on the relationship between science and religion.

Bibliography

Frankfurt, H. 2005. *On Bullshit*. Princeton NJ: Princeton University Press.

Habermas, J. 1985. *Die Neue Unübersichtlichkeit*. Frankfurt a.M.: Suhrkamp.

Habermas, J. 2005. *Zwischen Naturalismus und Religion*. Frankfurt a.M.: Suhrkamp.

Hempel, C.G. 1949. The Function of General Law in History. In H. Feigl & W. Sellars (eds.). *Readings in Philosophical Analysis*. New York: Appleton-Century-Crafts: 459-71.

Kant, I. 1962 [1781]. *Kritik der reinen Vernunft*. Hamburg: Meiner Verlag.

Merton, R. 1949. *Social Theory and Social Structure*. New York: Free Press.

Mill, J.S. 2010 [1859]. *On Liberty*. London: Penguin.

NAVF. 1976. *Vitenskapsteoretiske fag: rapport fra en konferanse om de vitenskapsteoretiske fags stilling i Norge, arrangert av Rådet for humanistisk forskning ved Norges allmennvitenskapelige forskningsråd, 29.-30. april 1975*. Oslo: NAVF.

Qutb, S. 1998 [1964]. *Milestones*. New Delhi: Islamic Book Service.

Popper, K. 1963. *Conjectures and Refutations*. London: Routledge.

Rawls, J. 1993. *Political Liberalism*. New York: Columbia University Press.

Shapin, S. 1996. *The Scientific Revolution*. Chicago: University of Chicago Press.

Skirbekk, G. 2007. A Crisis in the Humanities. In G. Skirbekk (ed.). *Timely Thoughts*, Lanham: University Press of America: 23-35.

Chapter 3: What is epistocracy?

Dimensions of knowledge-based rule

Cathrine Holst

The role of knowledge in political decision-making has been a central topic in political theory and social science for centuries. One central branch of these discussions has focused on the role of religious knowledge and authority in political rule and variations of "theocracy" or "rule of priests".[1] However, the central knowledge basis of a society or a political system is not necessarily of a religious kind. Arguably, in many contemporary societies the most crucial knowledge source is scientific and professional knowledge.

Certainly, to draw simple parallels between the role of secularized scientific and professional knowledge in today's liberal democracies with the role of religious knowledge in theocracies is to oversimplify complex relationships and to underestimate modern rationalization processes; to do so would be banal and very often incredibly anachronistic. On the other hand, to the extent that science and expertise are given unquestionable authority and their role in political rule is considered as "natural" or "God given", outside the scope of analysis and scrutiny, we may of course talk metaphorically – but stereotypically (insinuating that religious people are incapable of and uninterested in scrutinizing their religious beliefs) – of a "religious" state of mind or approach to scientific and professional authority. From this point of view, the discussions of this chapter are meant to be deeply "unreligious".

The more general discussion of the role of knowledge – be it religious or secular – goes back at least to Plato, who in *The Republic* recommended states to be run by philosopher kings. Recently, the topic has been re-introduced by David Estlund in his discussions of the legitimacy of "epistocracy", a "rule of the knowers" or knowledge-based rule, referring to the Greek word *episteme* (Estlund 2003, 2008).

The aim of this chapter is to contribute to the clarification of the meaning of "epistocracy" or a "rule of the knowers". We can think of epistocracy as an alternative to letting the wealthy rule (plutocracy), the property-owners (timocracy), a few

[1] Theocracy originates from Greek (*theos* means god) and refers literally to a "rule by gods or human incarnations of gods".

prominent families (oligarchy), the military (stratocracy), priests (theocracy),[2] or indeed to democracy, a "rule of the people". We can think of "epistocratic" and "epistocracy" along different dimensions. In the following the ambition is to introduce some central ideal type characteristics and variations of epistocracy – a proposal for an epistocracy typology.

This effort will profit from concepts and distinctions developed elsewhere, by philosophers and in different branches of empirical scholarship. A general typology of knowledge-based rule integrating relevant new and old insights from different academic fields and disciplines remains, however, to be spelled out. This fact makes sense given another fact – that of academic specialization – but it is also somewhat puzzling given the huge amount of scholarship on the role of science and expertise in political decision-making, and the fact that contemporary governments are knowledge-based in an immense variety of ways.

More concretely, in the following, eight dimensions of epistocracy will be elaborated:

1. the historical dimension (pre-modern or modern epistocracy?);
2. the organizational dimension (formal or informal epistocracy?);
3. the constitutional dimension (which constitution?);
4. the process dimension (where in the decision-making process?);
5. the substance dimension (decisions on what?);
6. the actor dimension (who are the knowers?);
7. the cognitive dimension (how rational?); and
8. the normative dimension (what justifies it?). A brief final part looks at the different analytical uses of a typology like this.

Concrete examples will be provided along the way; several of them from a European Union (EU) context. The EU is a unique and highly contested experiment in transnational governance with a myriad of epistocratic features, and thus provides a reservoir of examples and illustrations of the forms epistocracy may take in modern, democratic societies.

The historical dimension

Epistocracy will vary across cultural traditions, civilizations and historical epochs. A fundamental distinction is that between *pre-modern* and *modern* epistocracy. Paradigmatic contributions on what it means for societies to be "modern" and

[2] As already indicated, theocracy could also be regarded as a variant of epistocracy if we think of priest as "those with religious knowledge".

"modernized" (Giddens 1991, Habermas 1984, 1987, Parsons 1971), have been heavily criticized, both for their affirmative normative subtext (see for example Meehan 1995), and from an empirical point of view, for overlooking significant variation (Eisenstadt 2002). Still, it is possible to distinguish, at least roughly, between societies that are institutionally and normatively differentiated – societies in which the economy, the family, the political system, law, civil society, science, religion and art constitute relatively separate spheres that operate according to different standards and different "inner logics"[3] – and that have undergone what Talcott Parsons refers to as the three revolutions: the industrial, the democratic, and the meritocratic.

From the history of philosophy Plato's proposal for an exclusive rule of philosopher kings and Aristotle's more modest proposal for a rule of the wise are well-known examples of pre-modern epistocratic ideas. That is, these proposals are not only framed with reference to a "general acceptability requirement" (Estlund 2008) or a substitute that brings in a democratic standard, but also the modern normative horizon more generally; the institutional organization in which modern democratic standards are embedded and the context of the modern revolutions. Of course, from real history one could also pick from among a great variety of pre-modern systems of government with epistocratic characteristics, from the Roman Republic, to the Italian city-states, to the Enlightened absolutist regimes of Frederick the Great of Prussia, Catherine the Great of Russia or Joseph II of Austria (and then we would have limited ourselves to European history exclusively). What distinguishes these examples of pre-modern epistocracies from contemporary versions are single policies and institutional arrangements, the composition of knowers/rulers etc., but not least the fact that modern political rule takes place in modernized societies, in the sense described above, and within the modern horizon of normative and institutional expectations.

However, modernization also comes in degrees. If we return to the history of philosophy, John Stuart Mill's proposal of giving an extra vote to the educated in *Considerations on Representative Government* (1861) is arguably modern; Mill's horizon is in several respects that of a modern political thinker. At the same time many would deny that Mill's proposal is compatible with the contemporary democratic ethos. And modernization of course comes in degrees in real world cases, too. Consider, for example, the many ways in which religion is intertwined with law, morality and political rule in contemporary societies, and also the many debates that go on arguably within the parameters of the modern horizon with regard to

[3] "The project of modernity formulated in the eighteenth century by the philosophers of Enlightenment consisted in their efforts to develop objective science, universal morality and law, and autonomous art, according to their inner logic" (Habermas 1981). See also Skirbekk (2007).

what this horizon implies more specifically for the role of religion institutionally and normatively (Habermas 2008, Rawls 1997, Taylor 2008).

The organizational dimension

A rule of the knowers can be more or less formalized. In terms of formalization, constitutional epistocracy in one variant or the other is at one end of the scale, epistocratic laws and policies constitute an intermediate category, whereas social and cultural epistocracy in some version is at the other end. *Cultural epistocracy* refers to societies where respect for knowledge and knowers is considerable and many subscribe to the idea that decision-making must be knowledge-based and knowers must play a significant role in decision-making. Cultural epistocracy thus comes in degrees, and will also vary depending on which knowers and what kind of knowledge is recognized and valued. Sociologists have for example argued that formal academic degrees and merits are less valued in Scandinavian egalitarian culture than in other European countries such as in France and Germany; if by knowledge we mean academic knowledge, Norway is thus less of a cultural epistocracy than France. However, there are different kinds of knowledge; we can think of a cultural epistocracy that values moral knowledge, as in Plato's ideal state, or one that values technical expertise and technocratic knowledge.

Cultural epistocracy could also be a *social epistocracy* to a lesser or greater extent depending on whether the actual selection for ruling positions sociologically speaking is based on epistemic criteria (i.e. criteria referring to the position holders' knowledge), or retrospective sociological analysis shows that particular categories of knowers become rulers, even if this is not always explicitly prescribed or recognized.

Furthermore, cultural and social epistocracy can be more or less formalized in terms of *epistocratic policies and laws.* A system of government where recruitment prescriptions to ruling positions on different institutional arenas refer explicitly to education, cognitive competences etc. is more formally epistocratic than a system where this is a more implicit aspect of recruitment procedures. Compare, for example, a situation where laws and manuals pertaining to recruitment to public administration and courts contain requirements of a specific educational background (only candidates that have diplomas with distinctions are qualified, etc.), and a situation where this is the assumed framework. The degree of formalization may go hand in hand with a strengthening of social epistocracy – certain knowledge elites are reproduced because the recruitment procedures explicitly prescribe it – but not necessarily, if the epistemic criteria of recruitment are deeply internalized in recruiters to the extent that there is no need for formal codification.

The constitutional dimension

In its most formal version, epistocracy manifests itself in a written constitution and in laws and case-law with constitutional functions, as a *constitutional epistocracy*. The features of constitutional epistocracy can be discussed along different sub-dimensions; here the focus will be on the following three: constitutional norms on a radical-moderate continuum, degree of democratic accountability, and the unit regulated by constitutional norms on a state-non-state continuum.

First, constitutional norms could be more or less radically epistocratic with reference to an epistocracy-democracy scale. At one end of the scale, there is *radical* or *pure* constitutional epistocracy where only knowers are provided with basic civil and political rights. At the other end are *moderate* versions of constitutional epistocracy where all citizens are provided with basic civil and political rights, but where the constitution allows for or even prescribes epistocratic arrangements, such as a wide scope for experts and expertise arrangements in the agenda-setting phase or for judicial review. Depending on the amount and quality of epistocratic characteristics, it would at some point be more reasonable to talk about a constitutional democracy than a constitutional epistocracy.

Furthermore, a line could be drawn somewhere between moderate, but undemocratic constitutional epistocracies with reference to some democratic minimum standard on the one hand, and democratic epistocracies (constitutional epistocracies with democratic features) and epistocratic democracies (constitutional democracies with epistocratic features) on the other. One such minimum standard could be that of democratic accountability. A distinction could be made between moderate constitutional epistocracies where knowers with extra decision-making power are held accountable[4] – a *democratic* epistocracy (or an epistocratic democracy – depending on the amount and quality of democratic mechanisms) – and moderate, but *undemocratic* epistocracies where knowers with extra decision-making power are not formally or de facto held accountable or if they are, only in highly limited ways. This arguably goes to the heart of contemporary debates on democracy and democratic deficits. Consider, for example, recent European studies discussions of the accountability deficit of central EU institutions and procedures, such as co-decision and comitology (Curtin 2009, Eriksen 2009).

[4] In the sense that they are elected by all affected, by representatives elected by all affected, or appointed by someone who is elected, or appointed by someone who is appointed by someone who is elected, etc.

Finally, the unit of constitutional regulation can be more or less state-like. Typically, the units in question have been *states*. However, we could also think of constitutional norms for *less-than-states*, systems of governance that are lacking or have limited state functions. Once more, the EU could serve as example, but there is a web of international organizations and transnational networks with governance functions (Slaughter 2004).

The process dimension

Terms such as "political rule" or "political decision-making" can be interpreted narrowly as law-making, which in modern representative democracies is typically the task of parliaments. If so, the central question, from the point of view of epistocracy, is the extent to which the recruitment to and procedures of parliaments give certain groups of knowers extra power in the law-making process. One example is when the education level among parliamentarians is higher on average than the education level among citizens generally. Another example is the increasing significance of parliamentary staff and expert advisors employed by the political parties or working for parliamentary committees.

However, if in "political decision-making" we also include opinion formation, agenda setting (decisions on what?), and the making of option menus (decisions based on which options?) – *pre-decision* decision-making – and application and implementation of decisions – *post-decision* decision-making – the focus of investigators of epistocracy must be expanded. It would first require a study of opinion formation and agenda-setting processes in the public sphere and civil society. Are knowers privileged in these processes? If so, exactly which groups of knowers are privileged, and how are they privileged? This covers a wide range of phenomena, from powerful lobby groups' and corporations' reliance on high-level expertise in economics, law and science & technology, to the role of NGOs' expertise and counter-expertise, and the overrepresentation of people with academic training in the media. Secondly, the role of public administration, of governmental expert groups, and of agencies facilitating the administration, for agenda setting and option framing must be investigated. Which constellations of expertise inside and outside public administrations are given extra opportunities to shape the agenda? To return to the EU context, an example is the Commission's expanding expert group system, currently consisting of more than a thousand expert groups in different fields and policy areas (Sverdrup & Gornitzka 2010). Another example is the increasing role both in the EU and in the member states of knowledge-producing agencies, often connected to the program of so-called "evidence-based" policy-making (Nutley 2000).

Finally, in the post-decision part of decision-making, a study of epistocracy must once more investigate the role of bureaucrats and professionals in public administration – the extra power of administrative expertise constellations when laws and policies are to be implemented. In addition comes the role of courts and judges who interpret and apply law, be it in ordinary proceedings, or by means of judicial review where the legality of administrative and parliamentary decisions are tested. In an EU context this would call for a study of the European Court of Justice's decisions which have a direct effect in the member states and include a range of controversial issues. Consider, for example, the role of the ECJ in the field of gender equality and on the question of equal pay in particular. Article 119 of the Treaty of Rome stated the principle of equal pay for work of equal value, but it is the ECJ judges by means of their decisions in concrete cases that have defined what "equal value" means (Van der Vleuten 2007).

The substance dimension

Following Jürgen Habermas, we can distinguish analytically between pragmatic political decisions concerning factual and technical questions and strategic efficiency, political decisions that include substantive ethical considerations and political decisions pertaining to moral questions of justice and fairness (Habermas 1999). Accordingly, epistocracy will vary depending on whether knowers are given extra decision-making power over *factual/technical* questions, or also over *ethical/moral* questions. Consider, for example, the case of "gender experts" that are given an increased role in the field of gender equality policies following the development of state feminist machineries encouraged by the UN and now also prescribed for example by EU law (Lovenduski 2006). Arguably, the role of gender experts is not only to make recommendations with regard to policy efficiency based on empirical analyses; they are also given a privileged role when it comes to defining and operationalizing what gender equality ought to mean, and thus operate as "value experts".

Another distinction can be drawn between cases where questions left for the knower/rulers are *recognized as value-laden* and cases where these questions are *presented as value-neutral* and purely factual. An example is the role of economists and engineers in the development of environmental policies and technologies. To be sure, both economists and engineers offer expert advice on technical issues and policy efficiency, but their analyses also include more or less controversial risk and value assessments. Thus, economists and engineers very often intervene normatively in value discussions, even if this is not necessarily admitted or recognized.

Thirdly, questions can be more or less controversial and involve different levels of *risks and uncertainties* relatively independent of the fact/value distinction. That is, it is sometimes the case that factual questions are more straightforward and less controversial than value questions, in the sense that they are questions the relevant experts would agree on. However, there are also occasions where technical and scientific experts disagree strongly with regard to state of affairs, levels of risk and uncertainty, and policy efficiency. A recent example is the very different approaches among economists to the euro crisis.

Finally, political questions could be more or less *significant* in the sense that they make a difference – smaller or larger – to a smaller or larger group of citizens. A regime where many questions are left to democratic decision-making is not necessarily less epistocratic than a regime where fewer questions are left to democratic decision-making, if in the first regime a small group of knowers are privileged in decision-making and take the most significant decisions, whereas in the latter regime the group of privileged knowers constitutes a considerably larger part of the citizenry, but who concentrate their decision-making on smaller, secondary issues.

The actor dimension

Who are the knowers? Are they many or few? There is, first, a *quantitative* dimension to the question of who the knowers are, i.e. epistocracy would vary depending on the number of privileged knowers relative to the number of non-privileged/citizens. At one end of the scale, there are systems of governments that regard all citizens to be equally competent with regard to the issues touched upon in political-decision making (a rare example of a radical epistocracy that at the same time would qualify as populist democracy); at the other end are regimes where exclusive groups of knowers are given a privileged role.

However, the question of who the knowers are also has several qualitative dimensions. First, are knowers/rulers recruited on the basis of *theoretical knowledge* – acquired for example by means of an academic degree – or are they appointed as skilled *practitioners*? In the EU Commission's expert groups we find researchers and professors, but also public officials, professionals and NGO representatives with competence in and experience from the relevant policy field.

Secondly, are the knowers included in decision-making because they have a requisite *factual* knowledge and certain technical competences, or are those included rather considered to be *moral* experts, either because of particular practical skills or experiences or because of a particular academic training? An example of the first is when representatives from NGOs are included as experts because they are considered to be carriers of morally relevant knowledge and experiences as affected parties. An example of the second is perhaps when political philosophers are included in expert committees, presumably because by means of their training they have particular conceptual skills and experience in analyzing normative problems (Wolff 2011).

Thirdly, what exactly is the *content* of the experts' expertise? What is their disciplinary background? This may have decisive effects on the outcome. Consider for example an expert group reporting on Norway's relationship to the EU. A report authored by a group of lawyers would probably look very different from a report authored by political scientists or economists.

Fourthly, to what extent are the knowers highly qualified *specialists* within a particular, strictly defined field; to what extent are they *generalists* considered to have general insights and competences? Consider, for example, a case where the government asks an expert group to report on unemployment; causes of unemployment, effects, and possible policies addressing it. The government could then opt for experts who are high-ranking academics within labor economics or within other relevant, specialized fields. In reality, this is seldom what governments do.

What they very often do is opt for experts that are less specialized – relatively broadly oriented experienced social scientists that have general insights into social problems and research – but also that have been referred to as *interactional experts* (in contrast to *contributory experts*); i.e. they have the ability to interact and communicate across fields and disciplines, and with public official, politicians and even the public and civil society (Collins & Evans 2007).

The cognitive dimension

So far in this chapter, the meaning of epistocracy has been defined as the "rule of knowers" and "knowledge-based rule" interchangeably since knowers are those who know/have knowledge. However, a rule of the knowers may also be more or less knowledge-based depending on the extent to which knowers and others that take part in decision-making act in ways that maximizes epistemic or cognitive quality, i.e. delivers input to decisions or decision based on valid knowledge. This opens up firstly the question of the extent to which relevant actors are motivated to maximize their epistemic quality and to act on the basis of their best knowledge. Given this cognitive dimension of epistocracy, a rule where privileged knowers lack the motivation to behave like knowers may be less epistocratic than a rule where the privileged knowers are fewer, with fewer privileges, but are highly motivated to act as knowers.

Secondly, the epistemic quality of decisions is probably linked more or less closely to the deliberative qualities of decision-making processes, i.e. there is reason to believe that high-quality argumentation results in more rational outcomes than low-quality argumentation. In this respect, the literature on how to distinguish arguing and bargaining in political processes is also relevant for the study of (the cognitive dimension of) epistocracy, as are recent attempts to operationalize deliberative democracy, for example in terms of a Discourse Quality Index (Steiner *et al.* 2004). However, it should be emphasized that investigating deliberative quality within the parameters of democratic standards is not a task that completely overlaps with investigating deliberative qualities in political processes generally, i.e. relatively independent of such parameters. The latter is, however, what we would aim at in a study of epistocratic features of regimes and institutions. Accordingly, one should also look at possibly relevant contributions to the study and assessment of argumentation and deliberation coming from outside democratic theory, for example from philosophy of science and legal theory.

The normative dimension

A fundamental line can be drawn between those who defend *realist* justifications of epistocracy – like Niklas Luhmann (1984) who argues that we are functionally expertise-dependent so it cannot be otherwise; or Joseph Schumpeter (2005) who argues that people don't have the time and motivation to participate in politics – or *normative* justifications of epistocracy as "good" or "just".

Among the latter, one must distinguish between *substantive* justifications of epistocracy and justifications aimed at being *impartial.* Examples of the first may be a perfectionist justification – a justification of epistocracy with reference to ideas and assumptions about human flourishing, perfection and excellence and what facilitates it – or a justification with reference to cultural tradition ("respect for knowers and knowledge is a long-standing and valuable feature of our culture, this civilization etc."). Among justifications aiming at impartiality and justice, a central distinction is that between *input*-oriented or procedural justifications and *outcome*-oriented justifications. According to the first group of justifications, a political rule is legitimate to the extent that it is regulated by procedures with certain favorable qualities. According to the second group, a political rule is legitimate to the extent that it produces certain favorable outcomes.

One could furthermore distinguish between *democratic* and *non-democratic* procedural justifications and democratic and undemocratic outcome-oriented justifications. According to a democratic procedural justification, political institutions are legitimate, if they are regulated by democratic procedures (for example, in the sense that they are in accordance with the minimum standard of democratic accountability), whereas according to a non-democratic procedural justification, political institutions are legitimate if they are regulated by procedures with arguably favorable qualities (for example, procedures that maximize epistemic quality), even if they are not necessarily in accordance with standards of democratic accountability. A democratic outcome-oriented justification, on the other hand, regards political institutions as legitimate to the extent that they produce "democratic" outcomes, meaning here outcomes that all affected would subscribe to, be it here and now, or under some ideal conditions. A non-democratic outcome-oriented justification, finally, regards political institutions as legitimate to the extent that they produce arguably favorable outcomes (for example for the middle class, for the ruling elite or for some other group), even if they are not necessarily "democratic" in the sense that all affected would subscribe to them, be it here and now, or under ideal conditions.

Depending on the justificatory approach, epistocracy and different variants of epistocracy can be classified as more or less *legitimate* or *illegitimate*, as when Jürgen Habermas argues against the legitimacy of technocratic epistocracy on the basis of a democratic procedural justification, whereas David Estlund argues against Mill's unequal voting scheme on the basis of a democratic outcome-oriented justification. The choice of justification (should the justification standard be good outcomes or fair procedures – and if democratic, in what sense?) is a normative question that must be investigated separately, as the next step, i.e. if one aims also at distinguishing normatively between legitimate and illegitimate epistocracy, and not only descriptively, relative to different available standards.

It should be mentioned, finally, that there may be old and new interconnections between realist and normative justifications. It is well known that discussions about political ideals and norms must take into account certain "general facts about human society" and about modern society in particular (Rawls 1999: 119). Many would probably agree with John Rawls when he says that a theory of justice with relevance for modern societies must recognize, among other things, "the basic fact of pluralism": normative theorists cannot go about assuming that people of modern pluralist societies would ever come to share "comprehensive doctrines". A question that arises is whether we must also include what we might call a basic "fact of expertise" alongside "the basic fact of pluralism" and other basic facts normative theory must recognize. Should we regard modern societies' dependence on scientific and professional expertise as a "general fact" in Rawls's sense? If so, this fact must be reflected in one way or another in our normative ideas of legitimate political rule. The question, then, is how.

Why not epistocracy?

We could think of yet more dimensions in an epistocracy typology like this, for example a causal dimension (what explains epistocracy?) and other sub-dimensions (for example, on the substance dimension: are experts contributing to the decision-making process by means of describing or explaining something that has happened, immediate problem-solving or predicting long-term outcomes and effects?). Furthermore, a useful typology is always developed with an eye to what it is meant to be useful for. The epistocracy typology sketched here is meant to facilitate empirical work on epistocracy in a European Union context with a particular focus on the epistocracy-democracy relationship. Consequently, several of the epistocracy variables that have been identified in this chapter are also variables on an epistocracy-democracy scale, or outlined to provide information on democratic qualities and the epistocracy-democracy relationship. This also has implications for what is focused on as "knowledge". At the outset the understanding of knowledge underlying this chapter is wide and sociological: knowledge is that which is regarded as knowledge.[5] However, in practice, given the empirical focus, particular stress has been put on the role of scientific and professional knowledge.

Immediately, of course, urgent normative questions come to mind:[6] does not epistocracy and epistocratization constitute a threat to democracy, in the EU and elsewhere? To be sure, this epistocracy typology is intended to facilitate normatively oriented research that sheds light on this and similar questions – in the end. However, in research our ambition must be to make our normative assessments as specific and balanced as possible, without jumping to conclusions prematurely on the basis of general political intuitions. To put it bluntly: epistocracy is not inherently "wrong". Whether epistocratic arrangements are justified or unjustified normatively speaking, depends on the more specific characteristics of these arrangements and on the selected criteria of justification. Certain scores on epistocracy variables may refer to something totally unacceptable and illegitimate from a normative point of view (for example, an undemocratic constitutional epistocracy); other scores should perhaps not concern us that much, or should even leave us extremely relieved. Who would want policies, for example bio technology policies or gender equality policies, that were developed without any basis in scientific knowledge of bio technology and

[5] This of course does not imply that all definitions and criteria of knowledge are equally valid.

[6] The research project "Why not Epistocracy? Political Legitimacy and 'the fact of expertise'" (EPISTO 2012-2017) at ARENA – Centre for European Studies, University of Oslo, will undertake empirical studies of epistocracy in a European Union context (see http://www.sv.uio.no/arena/personer/vit/cathho/project_description_CH.html), but also contribute to discussions on normative assessment of epistocracy, philosophically, and connected to studies of EU institutions and arrangements.

modern societies' gender relationships? This reflects a commonly shared assumption, even among the harshest critics of expert rule and "technocracy" (Habermas 1999): that knowledge-based decision-making is acceptable and even a good thing if it is institutionalized adequately and legitimately, as well as democratically speaking, whatever that might mean more specifically (Collins & Evans 2007, Kitcher 2003, 2011, Longino 2003).

Bibliography

Collins, H. & R. Evans. 2007. *Rethinking Expertise.* Chicago: The University of Chicago Press.

Curtin, D. 2009. *Executive Power of the European Union.* Oxford: Oxford University Press.

Eisenstadt, S.N. 2002. *Multiple Modernities.* New Brunswick, N.J.: Transaction Publishers.

Eriksen, E. 2009. *The Unfinished Democratization of Europe.* Oxford: Oxford University Press.

Estlund, D. 2003. Why not Epistocracy? In N. Reshotko (ed.). *Desire, Identity and Existence. Essays in Honor of T. M. Penner.* Kelowna, B.C.: Academic Printing & Publishing. 53-69.

Estlund, D. 2008. *Democratic Authority. A Philosophical Framework.* Princeton NJ: Princeton University Press.

Giddens, A. 1991. *The Consequences of Modernity.* Stanford: Stanford University Press.

Habermas, J. 1981. Modernity versus Postmodernity. In *New German Critique* 22.

Habermas, J. 1984. *The Theory of Communicative Action. Volume I: Reason and Rationalization of Society.* London: Heinemann.

Habermas, J. 1987. *The Theory of Communicative Action. Volume II: Lifeworld and System: A Critique of Functionalist Reason.* London: Heinemann.

Habermas, J. 1999. *Between Facts and Norms. Contributions to a Discourse Theory of Law and Democracy.* Cambridge: The MIT Press.

Habermas, J. 2008. *Between Naturalism and Religion.* Cambridge: Polity Press.

Kitcher, P. 2003. *Science, Truth, and Democracy.* Oxford: Oxford University Press.

Kitcher, P. 2011. *Science in a Democratic Society.* New York: Prometheus Books.

Longino, H. 2003. *The Fate of Knowledge.* Princeton NJ: Princeton University Press.

Lovenduski, J. 2006. *State Feminism and Political Representation.* Cambridge: Cambridge University Press.

Luhmann, N. 1984. *Soziale Systeme: Grundriss einer allgemeinen Theorie.* Frankfurt a.M.: Suhrkamp.

Meehan, J. (ed.). 1995. *Feminists Read Habermas.* London: Routledge.

Nutley, S. M. *et al.* 2000. *What Works? Evidence-based Policy and Practice in Public Services.* Cambridge: Polity Press.

Parsons, T. 1971. *The System of Modern Societies.* Englewood Cliffs, N.J.: Prentice-Hall Inc.

Rawls, J. 1997. The Idea of Public Reason Revisited. In *The University of Chicago Law Review,* 64, (3): 765-807.

Rawls, J. 1999 [1971]. *A Theory of Justice.* Oxford: Oxford University Press.

Schumpeter, J.A. 2005 [1942]. *Capitalism, Socialism, and Democracy.* London: Routledge.
Skirbekk, G. 2007. *Timely Thoughts. Modern Challenges and Philosophical Responses.* Lanham, Maryland: University Press of America.
Slaughter, A.-M. 2004. *A New World Order.* Princeton NJ: Princeton University Press.
Steiner, J. *et al.* 2004. *Deliberative Politics in Action.* Cambridge: Cambridge University Press.
Sverdrup, U. & Å. Gornitzka. 2010. Enlightened Decision Making? The Role of Scientists in EU Governance. In *Politique européenne*, 32: 125-149.
Taylor, C. 2008. *A Secular Age.* Boston: Harvard University Press.
Van der Vleuten, A. 2007. *The Price of Gender Equality. Member States and Governance in the EU.* London: Ashgate.
Wolff, J. 2011. *Ethics & Public Policy. A Philosophical Inquiry.* London: Routledge.

Chapter 4: Doubt has been eliminated

Roger Strand

In a speech before the UN Commission on Sustainable Development, Gro Harlem Brundtland, the UN Secretary-General's Special Envoy on Climate Change, said:

> So what is it that is new today? What is new is that doubt has been eliminated. The report of the International Panel on Climate Change is clear. And so is the Stern report. It is irresponsible, reckless and deeply immoral to question the seriousness of the situation. The time for diagnosis is over. Now it is time to act (Brundtland 2007).

This chapter is not going to argue that Brundtland's statement indicates a religious or even close-to-religious belief in the ability of Science to speak Truth to Power. I am not going to propose that she expressed a religious or even dogmatic attitude by claiming that doubt has been eliminated in this specific case, i.e. the claim for anthropogenic climate change. What I will argue, however, is that Brundtland's statement as well as the curious ethics debate that followed in Norway – the story of which follows below – exemplify some of the peculiar challenges that arise when it is Science that is supposed to speak the Truth. Rather than framing these challenges in terms of the direct relationship between Science and Religion, I will borrow from Ragnar Fjelland (1985), who employed the Norwegian concept *livssyn* ("life philosophies"[7]) and analyzed how life philosophies may borrow authority from Science and Religion. I shall argue that Brundtland's statement indicates a belief in Science that has less to do with philosophy of science and more to do with life philosophy. I shall also criticize that type of belief, in a way that will be quite familiar to many contemporary (21st century) readers. However, even if the criticism may be trivial in intellectual terms, its acknowledgement in policy and governance still appears remote and unlikely.

Elimination of doubt and the ethos of science

"Doubt has been eliminated," according to Brundtland's statement, and the reason for this was that the "report of the International Panel on Climate Change" (AR4, the

[7] The literal translation of "livssyn" is "life view". One can find three translations of this term into English: (secular) "philosophy of life", "life philosophy" and "world view". The Norwegian Government appears to prefer "philosophy of life" in their official documents. In order to avoid the connotation to the Germanic philosophical tradition, I will use the slightly more awkward "life philosophy" in this chapter. "World-view" is in my opinion not an adequate translation because the concept of "livssyn" typically embraces and emphasizes the existential and moral dimension of human life.

Fourth Assessment reports of IPCC) and “the Stern report” (the Stern Review on the Economics of Climate Change) were “clear”. In other words, these publications were seen (by Brundtland) as carrying sufficient authority to be able to eliminate doubt in the competent and rational reader. It also seems obvious that their authority is based on their scientific character and credibility. Regardless of how Brundtland may or may not value scientific and political authority, *her* claim – coming from a UN special envoy, former Prime Minister of Norway and former Director General of WHO – about the elimination of doubt would appear quite ridiculous if she supported it merely by, say, the conclusion of a citizens’ consensus conference, the decision of a political party, or a declaration from an environmentalist NGO. What is noteworthy about the IPCC is that it is set up in order to ensure high scientific quality and legitimacy; and the Stern report is called the Stern report because of Sir Nicholas Stern’s position and academic reputation. Their authority does not lie in an acknowledgement of their moral superiority but in their presumed ability to *describe things as they are,* or to “Speak Truth to Power” if one is willing to accept this kind of jargon from science policy discourse. When Brundtland says that the reports are “clear”, it may mean they make it “clear” what needs to be done in terms of decisions and actions, but first and foremost the clarity is of a descriptive nature. It *is such that* anthropogenic climate change is upon us.

The problem, however, is that most contemporary philosophies of science – professional philosophies as well as the implicit and practiced ones that Søren Kjørup (1996) called “spontaneous philosophies” – would tend to grant continued discussion, open criticism and methodical doubt a central place among their ideals for scientific practice. Indeed, in the very center of twentieth-century expressions of belief in Enlightenment and Progress, thinkers such as Karl Popper and Robert K. Merton argued that a critical mind-set and the organization of skepticism are essential to Science and necessary for the maintenance of open and democratic societies. One may of course disagree with the Popper-Merton theses about the mutual dependencies between self-critical science and the open society, or about the empirical adequacy of describing scientific institutions and practices as the enactment of celebrated virtues of open-mindedness. Ever since Thomas Kuhn, the production of such disagreements has become an academic industry in itself. To express such disagreements, however, tends to be an act of distancing oneself from the official discourse on the Ethos of Science and how it is supposed to be embedded in scientific practice, taught in our universities, regulated by written and unwritten codes of conduct and employed in public decision-making. An argument in favor of the empirical description or normative prescription of “normal science” as a dogmatic enterprise rarely goes well with an argument for the supreme authority of the voice of Science in public decisions.

Unsurprisingly, Brundtland was indeed accused of anti-scientific attitudes in her embrace of IPCC and the Stern report; accused in a quite literal sense. In a great moment of late modern irony, the Norwegian Research Ethics Committee for Science and Technology (NENT) received a complaint in November 2009 about Brundtland's speech, arguing that it violated basic principles of research ethics: academic freedom, anti-dogmatism, organized skepticism.[8] NENT, a committee appointed by the Norwegian government and mandated by the Norwegian Act on Research Ethics, concluded in three parts. First, NENT clarified that it would not arbitrate on Brundtland's statement as such because she is not a researcher and the statement apparently did not intend to influence research practice. Secondly, however, NENT admitted the relationship between climate research and climate policy. A political speech, regardless of its intentions, may therefore indirectly and unwillingly influence scientific practice. Accordingly, NENT decided that the committee could and should comment on the content of Brundtland's speech. I have translated an excerpt of the comment:

> NENT finds it relevant to point out that accepted language use in scientific contexts differs from what one finds in the mentioned quote [*the introductory quote of this chapter, author's comment*]. Traditional academic norms allow and encourage doubt and critical questions. Doubt may in such contexts be well or ill founded, but not irresponsible and immoral by itself. In a situation of action, which then is not a purely scientific context, it may of course be irresponsible and immoral not to act, for example by maintaining doubt or criticism that one oneself finds poorly justified. It might be that Brundtland has this type of action in mind (NENT 2009).

NENT's statement goes on to mention precautionary principles as an example of principles of action designed to alleviate the tension between academic and political moral norms: "Such principles seek to justify political action while acknowledging scientific uncertainty and maintaining critical scientific debate" (*ibid.*).

In summary, NENT probably went as far as it could within its mandate, concluding that Brundtland's speech differed from "accepted language use in scientific contexts". In plain terms: her utterance violated the norms of the Ethos of Science. It would be a serious underestimation of actors at Brundtland's level, however, to think that her speech was a careless mistake or a result of ignorance. Of course she knew that it is "more scientific" to qualify statements; to appreciate the plurality of perspectives and expert opinions; to show awareness of the essential fallibility of scientific facts,

[8] The complaint made to NENT by cand. real. (≈ M. Sc.) Jan M. Döderlein as well as the Committee's reply is publicly available at http://www.etikkom.no. I should make clear that I was not a distant observer in this case, as I was a member of the NENT Committee at the time and participated in the drafting of the response to Döderlein.

theories and advice. Her task was a different one than being scientific: it was to argue for the supreme authority of science in order to combat doubts about the authority of the advice from IPCC and the Stern report.

The unscientific belief in science

General questions about the justification of the authority of perspectives and positions are very difficult and have received massive attention in (some would say *have plagued*) modern philosophy. In this chapter I shall not enter into what many would claim to be the more pertinent issue – namely how to justify one's special beliefs (some would say *comprehensive doctrines*) in a politically liberal society where others cannot be expected to share one's worldview or to endorse the same set of values or virtues. Rather, I will discuss some aspects of justification *from within* a perspective because I think it will shed light on the topic of this chapter, i.e. the relationship between science and life philosophies as exemplified by Brundtland's speech.

Justification of a comprehensive doctrine from within that doctrine can be anything from quite difficult to completely trivial. For instance, doctrines that postulate their own origin in revelations made by an omniscient, loving and truthful deity can have strong self-justificatory features. Of course we should believe God's words if they tell us that he is always right. Proponents of doctrines about the proper role and authority of science can choose a number of justificatory strategies. Sometimes, justificatory resources can be found within the perspective itself, as in the intriguing debate on the evidence for the utility of evidence-based practice in medicine. At other times, it has been found convenient to emphasize that justification in the last resort resides outside the perspective, as when Karl Popper points out the need to *decide* upon the role of rationality and the choice of critical rationalism. Critical rationalism is not consistent with claiming the *necessity* of its acceptance, if we are to believe Popper.

This is a relevant observation when discussing Brundtland's speech. There can be a scientific belief in Science – but if Science is defined epistemologically as fallible and praxeologically as an activity that embodies norms of doubt and self-criticism, the belief in Science cannot be too dogmatic and too hostile towards criticism raised against it without becoming unscientific. This problem is indeed what one may observe in the Brundtland quote. It claims not only that "doubt is eliminated" in this case but also that to raise further critical questions is immoral. It is very difficult not to see this as expressly unscientific and even at odds with the norms of the institution from which she borrows authority. The contradiction is perhaps not so important in itself. There is little reason to fear that climate scientists will become dogmatic just because Gro Harlem Brundtland made an unscientific claim about climate science. The interesting question is rather: if Science is not the source of authority for this type of belief in Science, what exactly is the source – as seen from within this type of perspective? Mere trust in IPCC and Stern and his team, however brilliant they may be, appears insufficient for such strong claims. If I am allowed to speculate, I would think that many observers would find it hard to trust such a complex and worldly endeavor as the IPCC *qua* institution to the degree that it eliminates doubt. Bearing in mind Brundtland's experience as former Head of Government and former Director General of WHO, it becomes even more counter-intuitive to imagine that she holds naïve beliefs about big international institutions. On this ground, "Doubt has been eliminated" appears less as an expression of reasoned trust in the worldly IPCC and more as an expression of faith in Science. What kind of phenomenon is that faith?

Livssyn – life philosophies

There is an abundance of potentially useful concepts for the problems I am discussing here. *Comprehensive doctrine* is one example. *Ideology, metaphysical position and worldview* are others. For instance, we could have followed in Georges Canguilhem's path and discussed how scientific concepts are exported and distorted into non-scientific contexts and become scientific ideologies. The point I wish to pursue, however, is not so much one of epistemology or political theory as one of "life philosophy" in Fjelland's definition. In the following, I shall discuss his analysis, as well as the Norwegian context into which it was introduccd.

Norway was and still is (2012) a confessional state. The Norwegian Constitution reads:

> All inhabitants of the Realm shall have the right to free exercise of their religion. The Evangelical-Lutheran religion shall remain the official religion of the State. The inhabitants professing it are bound to bring up their children in the same (Norwegian Constitution, Article 2).

This fact is not without practical complication in a modern welfare state. For instance, the State needs a way to distinguish between inhabitants who profess to the official religion and those who do not in order to keep track of public church taxes. A proportion of direct taxes collected from State Church members are directed to the State Church, whereas other, non-State religious communities receive the equivalent amount of public taxes corresponding to the number of their congregations. This principle is supposed to be one instantiation of the constitutional right to free exercise of religion. The economic and institutional dimension of "exercise of religion" is accordingly governed on the basis of membership in religious communities that exist in the designated official registry of such communities (that are approved by the State), and with reference to the Norwegian Act with the striking name *Lov om trudomssamfunn og ymist anna* ("Norwegian Act on Religious Communities, et cetera" (sic!)). In 1981, secular belief communities were mentioned by name in Norwegian legislation (and not only as "et cetera") in the Act Relating to Allocations to Belief-based Communities, and in public management the relevant category is now more often than not "tros- og livssynssamfunn", that is faith/religious and (secular) belief/life philosophy-based communities. A lot could be said about this, not least about the curious controversies that occasionally arise when the State decides not to approve of a particular community as worthy of being listed in the official state registry. For instance, the County Governor of Telemark withdrew the approval of "The Circle of Friends of Pi-ism" in 2006, referring to media news that indicated that the community was not a *serious* religious community[9] as they were "laughing" about their own approval.

In the obviously difficult work of demarcating what is a serious religious community and not, the Norwegian State supports itself with a definition made by its Ministry of Justice. The definition is quite comprehensive and requires that the (true) followers of a (serious) religion believe in a power or powers that determine fate, and that they direct their lives accordingly. Moreover, it "should" include fundamental concepts such as "holy", "revelation", "miracle", "sin" and "sacrifice" (Kirke-, utdannings- og forskningsdepartementet 2000, see p. 50 also for a self-critical discussion of this position). As for secular communities, it appears that Norwegian authorities never even tried to make a definition of their essence and instead opted for the more pragmatic solution of setting a lower limit of 500 members as a requirement for their approval. Still, the concept is not empty and appears to be endowed with an implicit expectation of seriousness and dignity, as exemplified by Norway's Prime Minister Jens Stoltenberg in a speech to Parliament 28 May 2010:

> President,
> Religion and life philosophy have always been an important part in the life of human beings. Our relationship to religion and life philosophy takes part in defining us both as individuals and as a society. *[He then proceeds to describe religious life, author's comment.]* [...] Others do not believe in a deity, but find a sense of belonging and guidance in a distinct life philosophy (Stoltenberg 2010).[10]

[9] See http://www.fylkesmannen.no/liste.aspx?m=5783&amid=1303323.

[10] Author's translation, which was anything but easy in this case. Stoltenberg uses the word "tydelig" – *deutlich* in German. I have translated it into "distinct", which perhaps exaggerates the association to rationalist philosophy in his mention of secular life philosophies.

Hence, the proper follower of a life philosophy – humanism and social humanism are the more visible ones in Norwegian public life – should perhaps not express *faith,* but at least "belonging" and "guidance" and definitely not just play around making silly jokes.

A close reading of Norwegian public documents still leaves one at a loss as to what a philosophy really is, however; one is rather left with a negative definition. It is almost a religion, but without religious beliefs: a type of Coca Cola Light or even Zero. At this point, Ragnar Fjelland's (1985) analysis may come to the rescue.

Fjelland argues that Kant's four questions of philosophy are the suitable point of departure for defining a life philosophy:

1. What can I know?
2. What ought I to do?
3. For what may I hope?
4. What is a human being?

Rather than reproducing Fjelland's argument, I shall apply his conclusion in the latter part of the chapter: one's particular answers to the three latter questions form a life philosophy. The answer to the first question – What can I know? – does not form an intrinsic part of the life philosophy, but may be central to its justification.

In this way, the concept of life philosophy is placed on a different level than religion and science. Religion and science may *provide* answers – inputs – to (or justifications for) the life philosophy, but they are not identical to the life philosophy. Fjelland shows how not only a religion such as Christianity but also a cosmology as found in Ancient Greek philosophy can provide answers to Kant's questions, and in this way justify particular life philosophies (from within the perspective itself, of course). Next, he argues that belief in science and progress can easily provide other answers to Kant's questions and in this way produce a science-based life philosophy. "Science-based" is a dangerous term in this respect, however. *Within* the proper domain of science, the quality of being "science-based" may endow a claim with superior epistemic authority. But Kant's questions are philosophical and not scientific ones; a categorical mistake happens if one believes that current biological theories can produce the unique and final truth about what is a human being, or if one uncritically embraces the technological imperative and concludes that we ought to develop and implement all technologies that can be delivered by science.

First or second modernity

Can one appropriately talk about life philosophies while discussing climate change? I think so. The issue of climate change cannot be separated from a number of huge questions about our responsibility for future generations, for global equity, for non-human species, and for our choice of lifestyles and therefore our values. I believe Brundtland and the author would agree on this point. In her speech, she was not trying to be a philosopher of science. She wanted to deliver a message about what is important and what we ought to do as societies and individuals.

Brundtland's speech borrows the answers to Kant's second and third question – what we ought to do and for what we may hope – from the IPCC and the Stern report. We should reduce emissions, and this can be believed to have a good effect. What a human being is, she in a way answered herself in the Brundtland report "Our Common Future", which not only states our responsibilities for future generations and across borders, but in that way also defines our roles and identity as intrinsically bound together on Planet Earth. Many would agree with her.

The problem appears with the relevance of Kant's first question: what can we know? By expressing her unscientific faith in Science, Brundtland undermines the authority of her life philosophy. It remains science-based, but no longer justified and endorsed by Science in its canonical expression. Nor is it supported by religion. Perhaps it could be supported by a cosmology of Simplicity – the Universe is, if not a Greek Harmonic Cosmos, at least such a simple place that even if scientists remain in methodological self-doubt, the knowledge they produce is in fact The Truth. This position is an intellectually vulnerable one, to the left of Religion and to the right of Science. Popper tried to find a way out if it; Paul Feyerabend spent most of his intellectual energy undressing and ridiculing it. Now, Brundtland's and our politicians' problem is not the presence of smart philosophers. The problem is that many 21st century citizens are endowed with critical skills and so little fear for authority that they no longer obey when leaders such as Brundtland say that doubt is eliminated and moreover immoral. Interpreted as an empirical statement, "Doubt has been eliminated" is quite simply false. One falls into ridicule and one's communication remains ineffective. The serious lack of agency that results, has been analyzed elsewhere (Rommetveit *et al.* 2010) and I shall not enter into political analysis here. I shall only recall Ulrich Beck's more general diagnosis of the class of political-environmental-human global problems that ever more appear to characterize our century:

> Wir leben in einer anderen Welt als in der, in der wir denken. Wir leben in der Welt des und, denken in Kategorien des entweder-oder. [...] Die stinknormale Weiter–und–weiter-Modernisierung hat einen Kluft zwischen Begriff und

> Wirklichkeit aufgerissen, die deshalb so schwer aufzuzeigen, zu benennen ist, weil die Uhren in den zentralen Begriffen stillgestellt sind. Die "Moderne" [...] ist in ihrem fortgeschrittenen Stadium zur terra incognita geworden, zu einer zivilisatorischen Wildnis, die wir kennen und nicht kennen, nicht begreifen können, weil das monopolistischen Denkmodell der Moderne, ihr Industriegesellschaftliches, industriekapitalistisches Selbstverständnis, im Zuge der verselbstständigten Modernisierung hoffnungslos veraltet ist (Beck 1993: 61 62).

NENT (2009) called for a Second Modernity type of approach: to admit that there is uncertainty in the climate models and still argue that this does not justify the lack of action; indeed the uncertainty may be a *reason* for precautionary action. Brundtland's problem is that she does not find strong enough power in a discourse of "and": science is telling us that the climate problem is really urgent AND Science may be wrong. Apparently unable to acknowledge Second Modernity, and no longer able to scare the people into silent obedience, leaders of 21st century democracies are simply left in deadlock.

Can we make constructive suggestions about how to get out of the deadlock? There are many already: epistemological ones (uncertainty and complexity management); political ones (deliberative democracy and a new social contract of science); legal ones (principles of precaution); ethical ones (eco-philosophy etc.). This short chapter shall end with a perhaps somewhat unusual suggestion: to think twice about our concepts of life philosophies. We have seen how even the Norwegian Prime Minister could not avoid quasi-religious concepts such as "belonging" and "guidance" when trying to describe life philosophies. As long as this submissive flavor prevails, secular life philosophies will remain too much a Coca Cola Light. One will be likely to fall into science-based but unscientific dogmatism and then into ridicule. Accordingly, I will end by making the claim for piecemeal, reflexive, self-critical and tentative life philosophies, allowing for the "and" of Beck and for doubts and smiles. A proper argumentation would require another chapter, or indeed a book series; still, let me forward the claim that a life philosophy of Beck's "and", fit for Second Modernity, would need to maintain hope in the absence of guarantees from God or from Science, and to see the questions of what we should do and what it is to be a human as deeply entangled (Funtowicz & Strand 2011) and relative to each other. I do so even if a community of such thinkers should not be found worthy by the County Governor of Telemark, Norway, to enter the appropriate State registry.

Bibliography

Beck, U. 1993. *Die Erfindung des Politischen: zu einer Theorie reflexiver Modernisierung.* Frankfurt a.M.: Suhrkamp.

Brundtland, G.H. 2007. Speech at the UN Commission on Sustainable Development. Retrieved from http://www.regjeringen.no/en/dep/ud/selected-topics/un/Brundtland_speech_CSD.html?id=465906.

Fjelland, R. 1985. *Vitenskap som livssyn? Livssynsspørsmål i teknikkens tidsalder.* Tromsø: Universitetsforlaget.

Funtowicz, S. & R. Strand 2011. Change and Commitment: Beyond Risk and Responsibility. In *Journal of Risk Research* 14: 995-1003.

Kirke-, utdannings- og forskningsdepartementet. 2000. *Tro og livssyn som grunnlag for tilskudd og offentlige funksjoner.* Rapport. Oslo: Kirke-, utdannings- og forskningsdepartementet. Retrieved from http://www.regjeringen.no/upload/KD/Vedlegg/Rapporter/troog050.pdf.

Kjørup, S. 1996. *Menneskevidenskaberne: problemer og traditioner i humanioras videnskabsteori.* Frederiksberg: Roskilde Universitetsforlag.

NENT 2009. Bør handle tross vitenskapelig usikkerhet. Retrieved from http://www.etikkom.no/no/Aktuelt/Aktuelt/Nyheter/2009/Bor-handle-tross-vitenskapelig-usikkerhet/.

The Norwegian Constitution. English translation. Retrieved from http://www.stortinget.no/en/In-English/About-the-Storting/The-Constitution/The-Constitution/.

Rommetveit, K., S. Funtowicz & R. Strand 2010. Knowledge, democracy and action in response to climate change. In R. Bhaskar, C. Frank, K.G. Høyer, P. Naess & J. Parker: *Interdisciplinarity and Climate change*. Abingdon: Routledge. 149-163.

Stoltenberg, J. 2010. Religionens plass i samfunnet. Retrieved from http://www.regjeringen.no/nb/dep/smk/aktuelt/taler_og_artikler/statsministeren/statsminister_jens_stoltenberg/2010/Religionens-plass-i-samfunnet.html?id=606053.

Chapter 5: The religious belief in rationality, science and democracy

Simen Andersen Øyen

> I continue to harbour a special prejudice against those who, in adopting this approach [neo-Kantian projects] imagine an ideal speech situation in which everyone (everyone?) would make the same moral and cognitive judgments. There are no moral or cognitive judgments which are not mediated by our concepts, and it seems to me that even our most apparently abstract concepts are historical through and through.
>
> *Skinner 2009*

In this chapter, I will argue that academic practice or disciplines understood as part of the Enlightenment ideals – here represented by Jürgen Habermas's communicative philosophy – must be contrasted with an understanding of ideals as normative regimes or imaginaries and how, then, academic practice must be reformulated as a historically and discursively based activity and set in the context of the late modern capitalistic society. I will focus attention on how the university as an institution is given a privileged position in society as a neutral provider of knowledge, expressed in the belief that technology and science will solve our fundamental problems, and as a democratic-moral corrective to the world outside – supposed to guarantee that political action is guided by knowledge, and that deeper and more encompassing scientific insights will provide a richer and truer picture of reality. The Enlightenment ideal, or understanding of modernity as a rationalizing process, has become one with Western, institutionalized science. I will discuss this problem complex by counterposing the image of modern academic institutions as a process of liberation through disciplinization and seen in light of our time's dogmas and fundamental beliefs: a free and open democratic polity, the free and just market economy and the autonomous pursuit of the truth called science (Wagner 1994). Common to these beliefs is a discourse of liberation which is tightly intertwined with a specific understanding of modernity.

These beliefs can be traced back to the pursuit of autonomy for scientific rituals and practices during the so-called scientific revolution, to the demand for self-determination in revolutions such as the French and American, and lastly, to the liberation of economic activities from the supervision and regulation of an absolutist and mercantilist state. One premise for a discourse connected to such an understanding of modernity is the construction of a subject capable of teleological action, self-controlling and autonomous. A version of such a discourse is put forth by Jürgen Habermas.

Moral imaginaries

The understanding of academic practices, and the ideals springing out of it, could rather be seen as normative regimes or moral imaginaries. The concept of moral imaginaries contains a belief in ideals, principles and norms as historically based and discursive. They constitute and decide the boundaries for what it is possible to legitimatize – controlling and regulating action, critique, decisions, deliberations, etc. The Western moral imaginary is for instance constituted by a series of historical events that have been especially central for the ethos of science, for instance the peace of Westphalia, the French and American revolutions, and the Holocaust (Sirnes & Øyen 2011). Ideas and norms are therefore not given, as in the classical rationalist approach, a form of independent existence as ideals and utopias, or as in the modified Habermasian variant, the status of regulative ideas. In such approaches there is a tendency towards an understanding of ideas as hypostatic. On the other hand, a concept of normative regimes will not reduce the normative level to superstructure, as is found in Marxist theory. The concept of moral imaginaries, or regimes, is related to Cornelius Castoriadis's concept (1987) of the social imaginary. This concept of moral imaginaries, however, must be understood in relation to how ideas constitute the limits of political and normative acts and expressions (Skinner 2009), and not only directed towards a semiotic and symbolic level in our understanding of the world. In this there is also a similarity to religion understood in a broad sense as a belief system: the idea of the Enlightenment as the historical realization of reason, the liberation of man and the equilibrium of the market form, together, a collective and social imaginary, overlapping on religion's function.

Even though ideas, ideals and norms may appear natural, they need to be placed in the context of anthropological analysis and a social analysis of modernity. The realization that we are not above or separated from an anthropological description of ourselves, through concepts such as myths, rituals, cultural patterns of action, etc., could lead to more comprehensive descriptions and understandings of the political and social relations we are part of. But if these relations are seen in light of an Enlightenment ideal and an understanding that everything is deliberatively available through rationality, then the anthropological aspects to our political, social and economic organization will be obscured. The conceptual and normative framework this ideal constitutes lacks an analysis of the genealogy of discourse, and the manner in which a discourse based in an ideal of Enlightenment itself contributes to forming the individuals' understanding of themselves and the social world.

The understanding of academic practice needs to start with the actual social, cultural and political practices of human beings. The practices which made knowledge possible must be studied in relation to knowledge itself (Lillebø 2007). For example, academic freedom is seen by many as a fundamental and ahistorical right. But it is, rather, a privilege springing from certain socio-economic relations in European history, particularly in how strong forces in the aristocracy desired to protect their scientists from the Church during the Renaissance. At that time, the Church controlled the majority of intellectual life and it was important to have a counterforce (Feyerabend 2011). Today the situation is reversed.

The ideals of religious freedom and the Enlightenment

Characteristic of the modern project is the desire to remove superstition and false consciousness. In the modernistic Enlightenment ideal – understood as both the belief in science and rationality – the history of science is read as an evolution in which steadily greater and better cultural and technological advances are made. It entails a conception of a political organization which, in the end, rests on the question of belief and the meta-story that democratization is a cumulative enlightenment process with clear and transparent distinctions between the public and private, the religious and secular. This understanding of modernity and liberal democracy shares some important features with religion. At its root there lies the thought that everyone – even Islamic fundamentalists or war-mongering American presidents – shall admit the "non-coercive coercion of the better argument." This belief alone is not problematic. The problem is rather that it presents itself as having a foundation in something universally human, ahistorical and precisely being a reaction to religious and existential approaches to the lifeworld – and thereby having a privileged epistemological position.

The faith in progress is not connected to a concept of God, but to the development of rationality. The modern project is built on an idea of the existence of a basis, which the truth rests upon, and which scientific methods can uncover. In Habermas, this basis is language and the development of steadily better approaches and procedures within each undifferentiated sphere of validity (Habermas 2004). Fundamental dimensions of experience within human existence are easily lost to view when the analysis of political processes is reduced to argumentation and rationality. Firstly, the Habermasian approach assumes that if we only know enough, we will also know what to do. But overall, there are other mechanisms – such as power interests, ideological structures of meaning, unconscious desires – which influence when decisions are made. Secondly, there is an assumption that one can always present things more true. The political and legal spheres have procedures for political formation of will which have become intersubjective: normative correctness is those claims which can be defended with general or universal arguments. In this lies an assumption that one can always be in better agreement and that one can coordinate actions more right. With such an idea, there follows a belief that the abolition and neutralization of power structures is possible. Discourse is the medium through which conflicting normative validity claims can be thematized and treated productively towards settlement in a consensual way. Effectualization or realization of discourse happens through a public, or a publically critical testing of validity claims in which all are assumed to have equal participation in a public rationality. This discourse is endowed by the scientific production of knowledge and arguments. Academic practice is therefore assumed to be a central part of the public sphere and a medium through which an inclusive opinion- and will-formation occurs in a more institutionalized sense (Habermas 1996). Public rationality is simultaneously defined as independent of context and ahistorical.

This type of understanding of the political, with an underlying uniform principle of a public rationality, rests upon a series of problematic assumptions. As Raymond Geuss (1981) points out, it is absurd to equate pre-dynastic Egyptians, French serfs from the ninth century, and Yanomami Indians from the beginning of the twentieth century within the view that they act correctly in as much as they are following a norm, which in an ideal speech situation would lead to universal consensus. Firstly, it is assumed that a liberal democracy is something all rational, consensus-seeking individuals are in agreement on, under certain idealized conditions which Habermas terms "ideal speech situation". This fits well with our own Western development. Here, the historical and cultural development becomes a series of stages which have all either brought about or anticipated our current political culmination: the liberal democracy. Secondly, politics are reduced to a certain region of society – the undifferentiated public sphere. Politics limit themselves to peaceful conversation and reasonable agreement, risking what the French philosopher Jacques Rancière (2005) points to as annulling the redundant subjects, reducing the people to a sum of social body parts and reducing political communities to a connection based on interests and rational desires. Thirdly, political processes are presented as rationalistic struggles over interests, in which individuals or groups are involved in conflicts that are defined by transparent parameters and rational arguments. This is built upon a legal philosophical fiction: the relations between political subjects are formed in an objective and comprehensible pattern, of transparency, and of the force of dialectic arguments. However, political plurality is not an autonomization of certain spheres, such as public rationality, but rather a variation of practices and publics which involve power relations, conflicts and irreconcilable interests.

The representation of indigenous peoples in the Western democratic institutions can be illustrative. Firstly, indigenous legal traditions are not recognized in the juridical system. These legal traditions are tied to myths, gender differences in which men have a privileged political position, and family hierarchies which cannot be translated into "egalitarian", rights-based deliberative processes. Secondly, there is a naïve belief that those outside our deliberative institutions and judicial system must be represented. The judicial system and the understanding of the subject as an individual bearer of rights are also not capable of recognizing collective rights. For example, collectively owned land areas (such as indigenous territories) do not qualify for financial credit, and in several countries in South America, such as Peru, it is possible to expropriate these territories if they are not commercially developed. For example, the sustenance and provision of the poorest 2/3 of the world's population is dependent on biological variation and local knowledge. This resource base is today threatened due to the patenting of poor people's plants and crops by scientists and Western firms who maintain they are "innovations" they themselves have created. The collective accumulation of knowledge by generations of local farmers, healers and indigenous peoples is not recognized within capitalist-technological market thinking. The West's post-industrial system with intellectual property rights forces onto the third world the WTO agreement, which concerns trade related intellectual property rights.[11] This dictate on intellectual property rights defines patents only as private property, never as collective rights. This means that all knowledge, all ideas, endeavors and habits which have appeared collectively – among village farmers, tribal peoples in the forests, or within local research collectives – are excluded. At the same time, intellectual property rights are recognized when knowledge or practice yields profit and capital accumulation, instead of meeting social needs (Shiva 2001). The common good is no longer acknowledged. In a Rawlsian pluralistic world, there exists no idea or ideals of "the good life" or "the good lives" – only liberal rights and the free market. Through reducing human knowledge to private property, the exchange of ideas is turned into theft and piracy.

A concept of a homogenous public rationality leads one in the direction of understanding politics as a moral and argumentatively technical field, where it is possible to attain a rational consensus through deliberation. This contributes to a depoliticization of social problems, making them the responsibility of professional politicians, experts, or bureaucrats to solve.[12] Separatist groups, political activists, those defined as "terrorists", indigenous peoples and subjects who stand outside the

[11] (TRIPS – Trade Related Intellectual Property Rights).

[12] It can also be said that this form of formation of opinion allows only advisory and expository interventions in the public sphere, and not that which has been termed parrhesia, in which one engages power through informed critique, and thereby incurs personal risk or the accusation of being too radical.

current discursive regime of recognition will not be heard. It is public rationality that defines the political subject, and this concept contains an inherent assumption that humans manage to exchange rational reasoning. The subject which operates in a social field as a fellow citizen is constituted as a self-reflective actor, who not only clarifies and strives for their own interests, but also actively defines and critically reflects over their own and others' desires against a background of competing motives and evaluations. The subject is constituted as an ideal, detached from a diverse cultural sphere of identification.

We will see, however, that this Enlightenment ideal has shifted in its contact with, especially, the market's functioning, though the idea that politics has its rational redemption has been retained. This shift can be seen as the Enlightenment ideal of today having been reformulated, so that scientification has been oriented towards the manifestation of technical projects. Sending a spaceship to Mars or developing a missile defense system for example has, and has had, a greater import for the USA's political agenda than eradicating illiteracy or poverty (Wilke 2009). Here a rationalistic Enlightenment ideal would contribute to, and underpin, an idea that "the human" and the social can be improved by objectivizing reconstructions so prediction and control would be possible. The latent functions of our contemporary form of political organization and mode of production are difficult to thematize, since the rationalistic ideal gives our late modern capitalistic democracy characteristics which are thought to be necessary. Hence, an Enlightenment ideal, as seen in Habermas, does not just conceal patterns of dominance and elite structures. It also risks becoming a theoretical support for a specific political regime – the liberalistic democracy.

The liberal dogma

The image of us all living in a society marked by freedom and democracy plays, without doubt, a decisive role in the Western political self-image. It is expressed in everything from various countries' constitutions to more popularized forms of commercial media. A striking example is when the prime minister of Norway, Jens Stoltenberg, after the terrorist attack on the 22nd of July, 2011, proclaimed that "our answer is more democracy and openness". After the terror attacks in New York on the 11th of September, 2001, George W. Bush said: "freedom itself was attacked this morning". The Swedish sociologist, Michael Carleheden (2009), would even go as far as to claim that the image of democracy in modern secularized societies has overtaken religion as "the opiate of the masses". Academic practice has a constitutive and maintaining significance for this image. Within a rationalist Enlightenment ideal it is assumed that academic practice is a democratic institution, precisely suited to delivering rational argumentation, guiding political action on the basis of knowledge, and holding a democratic-moral control function over society. From the ideal to reality, however, the road is long.

Late modern society is characterized by an extensive differentiation and specialization; different subsystems are separated, such as science, economy, law and politics. They are formed as special functional spheres and communication systems with accompanying demands for validity. Political intervention in the various subsystems demands knowledge on a given subsystem's specific function set. At the same time, these subsystems have rather a character of being systems in which knowledge can be seen as governing. This complexity is the primary reason why expertise is gaining increased political influence, not only through bureaucracy, but also in the execution of politics. Increasingly more areas are controlled by formal knowledge regimes, epistemic cultures and specialized communication codes (see Eriksen 2001, Luhmann 1992). At the same time as knowledge is institutionalized and demarcated as its own authority, it is forced to define its relationship to the rest of society, and with that appears a need for legitimation. Such legitimation acts as a democratic-moral corrective, finding expression in the ideal that knowledge should control and direct political decision-making.

This model is built on a positivistic scientific ideal which aims to obtain rational decisions on the basis of objective knowledge and is a form of "social engineering". The power of experts rests on neither citizenship nor representation, but on knowledge and the legitimation of a self-appointed scientific elite. This type of legitimation is based on the belief that social problems can be solved by accumulating empirical knowledge, and in the belief that the development of technology is a cumulative, uniform enlightenment project. In such cases, the rationalistic Enlightenment project will not provide an alternative, but rather risks underpinning and strengthening such displaced Enlightenment ideals as the belief in cumulative science and the attainment of technical projects. It does not, for instance, manage to formulate alternatives to a social imaginary which would not demand a high carbon footprint. In such a context, one can ask oneself whether academic practice, understood as an Enlightenment ideal, is not itself one of the constituting actors for overarching ideologies rather than a corrective to this system. Thinking, and its possible function as a corrective, is subordinated to the same economic-ideological paradigm that it is set the task to criticize. It must be seen in the context of the idea that the global capitalist economy has an unlimited capability to appropriate not just every attempt at new thinking, but also paradigms of knowledge and praxes that are not inherently subordinated to an economic-technological growth paradigm.

This can be seen in how universities are undergoing an encompassing adjustment to the market. Thinking is subordinated to the demand for so-called socially useful production of knowledge, teaching has become a commodity, etc. With this, the universities become education factories and corporate entities, with employed rather than elected leaders, a demand for the flow of students in which study points and feasibility are central, awards systems in which institutions are rewarded proportional to their publication rate, etc. The knowledge that is "produced" has value primarily in that it can be transferred to other undifferentiated spheres, be it the political-bureaucratic organizing of society or the development of new technologies and new products, thus being adapted to the meeting with state management and the market's imperatives. Today's invocation of academic ideas, such as "Bildung" and "public rationality" constitutes just as much a part of globalization's ideological-economic driving force as do alternatives to it (Jegerstedt 2011).

The technocracy, in the meeting with market imperatives, has further delivered the idea that social advancement is reached through increased economic production. Academic practice shall be organized as a market, through decentralization, privatization, and intensive thinking, being collectively known as New Public Management. Profession-political, commercial and paradigmatic thinking will govern. The transformation of social relations into market relation is, equivalent to expert's power, an attempt to solve certain of the governance problems which arise in complex societies. Control techniques such as goal setting, merit rewards and internal competition reflect this. These disciplinary fields are all constructed to control or form one or more limited aspects of the subject. Individualization can then be seen as a political control strategy (McRobbie 2009).

The effects of a capitalism which terminates all stabile relations between people and institutions are also clearly seen in academic practices. The new work environment is characterized by temporariness, risk, and "flexible" organization through, for example, the use of temporary employment in order to manage the continual reorganization in a shifting "information economy". On a structural level, it can be seen to concern new disciplinary and control techniques and new forms of structural inequality and power relations. At the same time, the social and structural risks in this type of system are individualized, reduced to personal inadequacies and a-politicized. The solution lies in the management of the self where various management theories can provide us with the indefinable qualities which are decisive for the development of different skills and abilities to better fit into today's structures of wage employment.

Who shall guard whom?

Precisely because the rationalistic tradition provides a shine of objectivity and creates expectations of rational, ahistorical and generally valid solutions to complex human problems, it is necessary to see it in a critical light. We all "know" that knowledge is more than the regurgitation of facts. However, because we "know" this, and because the academic practice endures an ongoing self-reflection, we are "free" to act within it and can legitimize it from a privileged epistemic position.[13] It is precisely ideals, the critical self-reflection, the rhetorical drills that separate this practice from other spheres of will- and opinion-formation or structures where meaning is constituted, and it is also just such a conception which divides the modern society from the "pre-modern" cultures in which the individuals think along lines of which they are not conscious: they are subject to discursive mechanisms they do not reflect upon.

The idea of academia's function as a channel for the public and free exchange of ideas through open and transparent argumentation, as an independent conveyer of knowledge, and as a control element over other differentiated spheres must be reformulated in light of a historical and discursive understanding. It is not ideas and ideals which are primary, but praxes as historical, discursive processes where one balances materialistic structures with cultural and normative regimes, individual expression with collective complexes. This means that one will not see the history of knowledge on a level of truth, but rather investigate which relations truth is constituted in.[14] Thus, the academic practice and the associated ideals can be understood as a normative regime, as an ideological and cultural construction that manifests itself through a system of different institutions and intervenes in people's lives through institutional networks, and as a product of society's various historical, social and political processes. This means that the academic practice does not necessarily inhabit a privileged position as a democratic corrective, but should be monitored with the same suspicion as all other activities in society. Such control cannot satisfactorily occur through ongoing self-reflection or researchers guarding themselves, as in various centers and committees such as the Center for the Study

[13] For example, in every number of the Norwegian Social Anthropologic Journal over the last eight years, there has been at least one article that explicitly treated one or more of the central scientific theoretical questions in the Western history of ideas: objectivity, generalization, representation, advocacy, etc. The rhetorical drills around method that every dissertation in sociology must include have an equivalent status: as long as science self-reflects on its methodological and scientific theoretical problems, it can choose to address that which separates it from other structures where meaning is constituted.

[14] For example, the psychiatric evaluation, released in the fall of 2011, which declared the perpetrator of the 77 murders on the 22nd of July of that year in Oslo and on Utøya as "criminally insane", has served to strangify, defamiliarize and historicize psychiatry as a scientific practice. The discussion since this report has been a reminder that knowledge does not arise in isolation, but rather exists in a social and cultural practice in which diagnoses must be interpreted just as much as concepts defining social position and normality at a given historical moment.

of the Sciences and the Humanities or The National Committee for Research Ethics in Science and Technology. This doesn't mean that Norwegian researchers should have anthropologists from the Third World come and study us. Neither does it mean that we need a variant of the Chinese Cultural Revolution. But it does mean that the privileged position which is ascribed academic practice in light of an idea on Enlightenment must not only be critically investigated. It must also be seen as a legitimization of the emergence of a hegemonic capitalist-technological growth paradigm where it feeds the political arena and market with a – rightfully enough, misrepresented – Enlightenment ideal. As such, the position of academic practice must be seen as part of one of the most imperative problems of the 21st century.

Bibliography

Carleheden, M. 2009. Är demokrati möjligt i komplexa samhällen. In *Fronesis* 29-30: 8-25.

Castoriadis, C. 1987. *The Imaginary Institution of Society*. Cambridge: Polity Press.

Eriksen, E. 2001. *Demokratiets sorte hull: om spenningen mellom fag og politikk i velferdsstaten*. Oslo: Abstrakt Forlag.

Feyerabend, P. 2011. *The Tyranny of Science*. Cambridge: Polity Press.

Geuss, R. 1981. *The Idea of a Critical Theory: Habermas and the Frankfurt school*. Cambridge: Cambridge University Press.

Habermas, J. 2004. *Teorien om den kommunikative handlen*. Aalborg: Aalborg Universitetsforlag.

Habermas, J. 1996. *Between Facts and Norms*. Cambridge, Mass: MIT Press.

Jegerstedt, K. 2011. "Teaching in the End Times": Humaniora og kunnskapsproduksjon i globaliseringens tidsalder. In S. Øyen *et al.* (eds.). *Humanioras fremtid*. Oslo: Cappelen Akademiske Forlag. 113-137.

Lillebø, J. 2007. *Vitskapleg rasjonalitet mellom sanning og historie: Vitskapsteoretiske refleksjonar over den vitskaplege forgjengaren i immunologiens historie*. Master's thesis. Bergen: Department of Philosophy, University of Bergen.

Luhmann, N. 1992. *Tillid*. København: Hans Reitzels forlag.

McRobbie, A. 2009. Alla är kreativa. In *Fronesis* 31: 93-116.

Rancière, J. 2005. Tio teser om politik. In *Fronesis* 19-20: 96-114.

Sirnes, T. & S. Øyen. 2011. *Moral Imaginaries*. Strategic document. Bergen: Centre for the Study of the Sciences and the Humanities, University of Bergen.

Skinner, Q. 2009. *Politik og historie*. København: Hans Reitzels Forlag.

Shiva, V. 2001. Världen vid klippans rand. In *Fronesis* 6-7: 170-190.

Wagner, P. 1994. *A Sociology of Modernity: Liberty and Discipline*. London: Routledge.

Wilke, H. 2009. Upplysning genom expertis og expertsystem. In *Fronesis* 29-30: 8-25.

Chapter 6: Psychology as science or psychology as religion

Historical presumptions and consequences for the present

Ole Jacob Madsen

> Psychology and religion will perhaps in the future need to be seen as two different but related ideological frames for constructing images of the self.
>
> *Carrette 2001*

Hermann Ebbinghaus (1908: 9) once famously remarked that: "Psychology has a long past, but only a short history". The short history tells us that Wilhelm Wundt founded modern psychology as an independent science when he established the first experimental research laboratory in Leipzig in 1879 devoted to the study of basic human reactions like sensations, attention and perception (Boring 1957). Psychology's brief, yet highly successful (his-)story is well-known as this lesson is taught at most introductory courses in psychology around the world. However, psychology's long past usually remains less illuminated, or if told, presents the listener with a narrative where modern day psychology is the unremitting highpoint of Western pre-scientific conceptions like Aristotle's rejection of Plato's ideas of the soul (Parker 2007). Even so, the recurring idea of the present age as postmodern, and psychology as a project of modernity, means that the science of psychology might be out of touch with the current age (Kvale 1992). One of the many implications of postmodernity was a shift from the sole study of the interior individual psyche to the practical repercussions of psychological knowledge in society, including epistemological, ethical and political implications (Kvale 1992). The postmodern rupture in confidence in Western science means that psychology can just as easily be understood as a substitute for religion in providing the fundamental guidelines for life. Yet, the religious roots and assumptions of psychology are seldom explored in full. However, Steinar Kvale (2003) pointed out that a number of the pioneers in psychology from Wundt and onwards to Carl Jung and Carl Rogers were sons of ministers, and religious father-son conflicts in fact had an important influence on the psychology they later preached. For instance, Jean Piaget sought to develop a science of psychology that was consistent with his religious beliefs – such as his conviction that all living organisms developed towards equilibrium with God as the ideal equilibrium. But, as Kvale (2003) comments, history books in psychology tend to emphasize Piaget's biological inspirations, and systematically fail to mention Piaget's religious ambitions (including two whole books on the topic where Piaget lays out the program for his life's work). There are in the present obvious reasons for the discipline of psychology to ignore postmodern skepticism and keep retelling and enhancing the official tale of psychology as an experimental branch of the natural sciences, e.g. in terms of future claims to external funding of research (Pawlik & Rosenzweig 2000). Nonetheless, even if we situate psychology among the "hard sciences" this positioning should be followed up by reflections on psychology's theoretical and ethical foundations. As one of "the founding fathers", William James, famously assessed over a century ago:

> When, then, we talk of "psychology as natural science" we must not assume that means a sort of psychology that stands at last on solid ground. It means just the reverse; it means a psychology particularly fragile, and into which the waters of metaphysical criticism leak at every joint (1985: xiii).

In this book chapter I will make a small addition to the important ongoing reflection on psychology's foundations. The question I propose is whether there actually are potential benefits of a view of psychology as religion and a closer alliance between the studies of psychology and religion. I will pursue the notion of psychology as a religion, or at the very least psychology as sharing some important features with religion, and pose the question whether an understanding of contemporary psychology as religion actually holds some ethical advantages over the belief in psychology as a pure science in today's mounting "therapeutic culture". The term therapeutic culture (or therapeutic ethos) is now a widely used sociological concept of how psychology is currently leaving an imprint on contemporary society (Wright 2011). For instance, Svend Brinkmann (2008) maintains that psychology has severely influenced the social imaginary of the West, and "universal" psychological ideals of individual self-realization now serve to reproduce social pathologies like identity problems and depression. Yet, most professional psychologists refuse to recognize the wider cultural implication of their increased presence in late modern life (Prilleltensky 1997). "Psychology as religion" might equip professionals with an improved understanding of "small-p psychology's"[15] present expansion, and a stronger sense of ethical responsibility towards the therapeutic culture.

From Protestantism to therapy

Whenever the question "Is science a religion?" emerges on the critical horizon (Dawkins 1997, Einstein 1949), psychology in many ways stands as an obvious candidate for anyone who wants to make the claim that modern science has come to replace or taken on the status of a world religion like Christianity. This holds

[15] Large-P psychology refers to the formal, institutionalized discipline of psychology with its academic departments, journals, organizations, etc., whereas small-p psychology refers to psychology in general and takes the form of everyday psychology through which people make sense of their lives (Pickren & Rutherford 2010).

especially true since psychology offers a rationale to human suffering, which has traditionally been among the most vital aspects of all world religions (Weber 1970). Kvale (2003) maintains that both traditional religion and modern psychology are equals in the sense that they provide a certain worldview, with a set of visions of the good life as well as concepts and techniques that help mankind in his quest to organize both the interior life of the psyche and the exterior life-world of the social sphere. Kvale (2003) traces the historical roots of modern psychology back to Protestantism in the sixteenth century, in particular its key characteristics like individualization and the construction of the inner person, and successive remedies in truth-seeking and contemplation and confession in pastoral care. Hence, modern psychology is comparable as a secular replacement to the Christianity that ruled the medieval age, after God was interchanged by man as the fundamental center of the universe during the Renaissance. Yet, although this historical shift away from a Christian worldview towards a therapeutic *Weltanschauung* is conceivable as a consequence of Western modernization, such an account remains disputed for several reasons.

The turning away from religion

The great classic in psychological analysis of religious experience, William James (1982: 515) maintained in his *Varieties of Religious Experience: A Study of the Human Nature* (1902) that religious experience may bring man into an altogether other experimental world of consciousness than the sensible and understandable world of things: "The whole drift of my education goes to persuade me that the world of our present consciousness is only one out of many worlds of consciousness that exist, and that those other worlds must contain experience which have a meaning for our life also." Hence, it is possible to argue with James that a complete rationalized, scientific world leaves religion's radical potential in shattering the merely sensible and understandable world of mankind unused (Freeman 2001). I will not dwell further on the promise of the more esoteric possibilities of religious experience here, but simply note that it looks as if the communitarian critical reception[16] that followed in the footsteps of modern psychology's unfolding in Western civilization throughout the 21st century, at least in parts have in common the same apparent religious notion that psychology regrettably drives a wedge between man and society (Illouz 2008). Still, the communitarian criticism of individualistic psychology is more secularized than James believes, as it owes much to Émile Durkheim's vision of "religion" as something eminently social. In his classic study *The Elementary Forms of Religious Life* Durkheim (1971: 419) spells out that: "If religion has given birth to all that is

[16] Communitarianism is a philosophical school that emphasizes the connection between the individual and the community. This critique of modernity has frequently reprehended psychology for promoting an ideal of the atomistic individual that only reinforces the ills it claims to heal (see for instance Bellah 2008).

essential in society, it is because the idea of society is the soul of religion." Therefore, modern psychology frequently stands indicted as something disruptive of religion, here defined as a crucial part of the social system. For instance, the arguably most influential single work on the therapeutic culture, American Freud scholar Philip Rieff's *The Triumph of the Therapeutic: Uses of Faith after Freud* (1987), takes a typical Durkheimian point of departure in postulating a crack in Western culture where the therapeutic ethos comes to overthrow the old Christian worldview. The result is an emerging therapeutic cultural movement which fails in the most fundamental function of culture in directing man outwards from himself and integrating the self in his communal and symbolic surroundings (Rieff 1987). Despite the fact that Rieff (2006: 13) is famous for his doomed forewarning that "No culture has ever preserved itself where it is not a registration of sacred order", his ideas of religion like Durkheim remain secular and anthropological in the making. Now, Rieff's wide influence on the scholarly field that followed devoted to studies of the therapeutic culture has been reproached for reproducing a secular bias that is out of step with the recent "return of religion" (cf. Casanova 2011). For instance, historian Christopher Loss argues that: "Scholars of the therapeutic, therefore, must stop acting as though religion does not matter when it clearly does" (2002: 71). Loss therefore calls for closer comparisons between the therapeutic ethos and other possible competing or complementary codes of moral understanding currently at work in Western culture: the question is what can we gain by comparing psychology with religion? All the same, prominent Christian scholars have in past decades endorsed a more theological and ill-disposed articulated basis against psychology's recent rise to power.

Psychology as religion

The explicit notion "Psychology as religion" was commonly associated in the 1950s, 1960s and 1970s with a certain fundamental critique of psychology, as modern psychology allegedly contained ambitions to replace religion by making claims to the ultimate purpose of life (Carrette 2001). Thinkers like the Christian theologian Paul Tillich and Catholic psychologist Paul Vitz launched a "Psychology as religion" condemnation of psychology for signifying a kind of hubris where it transcends its role as an empirical science and takes on the character of a myth, as the focal framework for the whole of Western culture (Bregman 2001). Tillich's (1957) position is that if scientific psychology, like Sigmund Freud and many of his followers, attack faith they are guilty of representing another kind of faith themselves, overstepping the scientific analysis. For Tillich (1957: 126-127) faith is man's ultimate concern which never can be undercut by modern science or any kind of philosophy: "Faith stands upon itself and justifies itself against those who attack it, because they can attack it only in the name of another faith. It is the triumph of the dynamics of faith that any denial of faith is itself an expression of faith, of an ultimate concern." However, when Vitz addresses the same issue two decades on it is as if psychology has now taken on the prominence of a religion. Vitz opens his book *Psychology as Religion: The Cult of Self-worship* (first published in 1977) with the following appeal to his readers:

> This book is for the reader interested in a critique of modern psychology – the reader who knows, perhaps only intuitively, that psychology has become more a sentiment than a science and is now part of the problem of modern life rather than part of its resolution (Vitz 1991: 9).

By portraying some of the influential self-theorists from this period like Erich Fromm, Carl Rogers, Abraham Maslow and Rollo May, Vitz (*ibid.* 37) advances to the conclusion that "self-theory is a widely popular, secular, and humanistic ideology or "religion," not a branch of science". Vitz (*ibid.*) bases his radical conclusion on Fromm's own definition of religion: "any system of thought and action shared by a group which gives the individual a frame of orientation and an object of devotion." Still, the difference between Christianity and psychology's underlying assumptions is vast according to his analysis. Vitz (*ibid.* 91) maintains that the relentless search and glorification of the self is at direct cross-purposes with the traditional Christian injunction to lose the self: "For the Christian the self is the problem, not the potential paradise." Vitz traces the problem of modern self-psychology back to the ancient conflicts Christianity faced with Stoicism and other sophisticated Graeco-Roman philosophical and ethical systems that conduct self-worship and self-realization, which in Christian terms is simple idolatry stemming from the old human motive that is egotism. There is also a profound conflict between Christianity and psychology regarding the nature of suffering, while only the former acknowledges evil, pain and ultimately death as a fact of life, but also as a source of transformation into transcendence (*ibid.*). Nonetheless, Vitz remains hopeful that the millions of people living their lives under the spell of science and psychology in the post-Christian era out of necessity or simply boredom will again seek to return to the arms of God.

Interestingly, Vitz at times proposes what can be labeled a natural-scientific critique of psychology (Teo 2005): self-theory psychology is a popular secular substitute for religion as it is not a science, because it fails to successfully scientifically operationalize its fundamental concepts like "the self" (Vitz 1991). Instead concepts like "self-actualization" and "self-realization" becomes vaguer, further removed from their original conception and seemingly more based on faith in charismatic predecessors like Maslow and Rogers. Vitz (1995) would later return to this type of critique in the 1990s when he gave a talk about perhaps the most evident concept of the globalized, therapeutic culture today – self-esteem – which Vitz concisely takes apart as a fundamentally muddled and confused concept and incapable of scientifically predicating behavior. Particular his critique of self-esteem illustrates how Vitz in a sense is both right and wrong at the same time in his view of psychology as religion. Even though Vitz has a profound understanding of all the good reasons for abandoning "self-esteem" as a useful concept, he simultaneously fails to see that this same ambiguity at the heart of the therapeutic movement is perhaps the reason for its global success. Self-esteem is loose and flexible enough to be stretched and applied almost within any sphere from education to parenting advice to self-help manuals to clinical psychotherapy. Instead Vitz remains devoted to the standard of psychology as a real science on the one hand (he mentions psychoanalysis and psychiatry, studies of animal behavior from biologists and ethologists and the research of experimental psychologists as exceptions), and his Christian faith on the other. This leaves us with a split between proper science and religion, which becomes a private matter of faith. Without going into the difficult aspects of liberal democracy and the proper role of religion as something fundamentally public or private, I will for completely different reasons propose that psychology should be situated somewhere in-between the polarities Vitz aspires to.

Psychology as religion reconsidered

The Christian critique of psychology, which maintains that psychology functions as a substitute for faith, will say that psychology cannot support man in an ever more chaotic modernity. This critique is sometimes confused as to the essence of psychology. Psychology's tremendous success in establishing itself as a vital reference point in a growing number of niches, including Protestant religion (cf. Smith & Denton 2005), suggests that psychology must be recognized as a faith of global reach. Vitz and fellow Christian believers are right in stating that psychology has become a religion, "a form of secular humanism", yet are mistaken when they complain that psychology is "based on worship of the self" (Vitz 1991: 9). "Psychology as a science" should be contested, but not because it has become a cult of self-worship, as this moralizing critique completely overlooks how the advanced liberal democracies in the West are founded on self-government (Rose 1996). From this perspective, the concept of "self-esteem" must be understood, not simply as egotism, but as a self-governing human technology upon which man can operate on himself so that the therapist, doctor or police do not have to (Cruikshank 1996). Psychology has succeeded, not because it is a religion of self-worship, but because it is an effective modern religion of self-governance that through commonly available therapeutic concepts like "self-realization" and "self-esteem" help people independently manage and cope with their everyday problems (Illouz 2008). Despite this forgiving and pragmatic view of psychology as religion, there are still important critical tasks ahead, namely to analyze the religious elements of what constitutes what has been called "the governing of the present" (cf. Miller & Rose 2009).

Professor of religion and culture Jeremy Carrette (2001: 121) has taken on this task as he envisions three major responsibilities for a future psychology of religion that takes the critical rethinking of psychology seriously: "first, an examination of the social and historical roots of human image construction and identity; second, an exploration of the religious ideas that infiltrate into psychology, and third, a critical assessment of the models of human beings provided by psychology." Through a reconstruction of psychology and religion Carrette (2001) maintains that religion and psychology must be seen as an interconnected discourse within the framework of what Michel Foucault (1978) called *governmentality*. The advantage of this re-positioning is that the traditional dichotomies between "religion" and "psychology", "body" and "mind", and "individual" and "social" are suspended, according to Carrette. Thus the future of the psychology of religion is founded on a cultural psychology which both recognizes the diversity of human image construction and modes of introspection, and ideally makes possible an ethical and critical analysis of the dominating ideas presently at work outside and inside the subject (Carrette 2001). It is also worth mentioning that Carrette takes a comparable position like the previously mentioned Loss: the question we must ask ourselves is what are the possible benefits of a closer alliance between the studies of psychology and religion?

Going back to the roots

Carrette reserves the scope of his ambition mainly for the subfield of the future study of psychology of religion. However, if we stick with the "Psychology as religion" approach, this means that the reconfiguration of psychology of religion could apply to the whole field of psychology. In fact, a critical examination of the kinds of human beings produced in the present may not actually be such a radical departure from what some of psychology's pioneers once envisioned. If we briefly return to general psychology's short history, Wundt himself divided psychology into the experimental branch that could serve the study of basic human operations like sensations, attention and perception and what he called *Völkerpsychologie* which was devoted to the study of the higher human processes like language, historically evolved forms and cultural artifacts. This part of psychology however never succeeded in exerting the same kind of influence on the shape of modern psychology as the experimental branch did throughout the 21st century (Danziger 1990). The path not taken through *Völkerpsychologie* and its successor cultural psychology, could have designated a place for psychology somewhere between the natural sciences and history where different groups of individuals' mental capacity is a dynamic ever-changing state

(Diriwächter 2004). Psychology as a cultural founded psychology would then easily interpret itself as a certain kind of religion, more specifically as a theoretical worldview which necessarily leaves an imprint on the ways an individual perceives himself (introspection) and perceives the world around him. The founder of the University of Berlin, Wilhelm von Humboldt, was also vital in the development of *Völkerpsychologie* (*ibid.*), as he maintained that every language contains its unique form, based on a particular Weltanschauung, which means that humans simply by being born into a common language community are immediately exposed to a certain relationship with the world. Alas, both psychology's own subject matter – the human psyche – and psychology as a helpful tool for the mind's study of itself is reasonably comparable to a religion that according to Fromm's previously stated definition is a system of collective thought that gives the individual a frame of reference for his orientation in the world.

Returning now to the overall question whether psychology can reasonably be considered a science or a religion leaves us with several options. (1) *Psychology is a science and not a religion.* This is the official and most commonly held view, where modern psychology fits in the bigger picture of a general progression of modern science evolving from the Age of Enlightenment. (2) *Psychology is a religion and not a science.* This position is held by the Christian critique of psychology as represented in this chapter by Paul Vitz.[17] Although Vitz holds important insights into the radical nature of how self-psychology alters culture, his critique is still caught up in a limited moral view where self-psychology becomes the latest expression of a cultural vice of selfishness. I argue that this standpoint is not justified, as it among other things gravely neglects the subjective turn in Western society (cf. Taylor 1992). This leaves us with a final option: (3) *Psychology is a science and psychology is a religion.*[18] This viewpoint consolidates and recognizes that psychology, as a natural science, can lead us to important new insights in, for example, the neurochemical origin of severe mental disorders like schizophrenia and bipolar disorder which would benefit a large share of the population. This duplicate view would simultaneously place modern psychology in the middle of a cultural debate on the role of the individual and society without pretending to hold a neutral position concealed by its status as a scientific enterprise producing value-free knowledge. From the standpoint of ethics, and in particular *area ethics* (cf. Nafstad 2008), psychology as a science and psychology as a religion, would represent a leap forward from the present situation which is characterized by outdated professional ethics based on an idea of social responsibility

[17] Yet, as previously mentioned, Vitz still acknowledges some parts of psychology as science.

[18] I deliberately leave out the option that psychology is neither a science nor a religion. From a strictly logical point of view this is of course a perfectly viable outcome, if psychology fails to meet the inclusion criteria either way. Yet, this possibility exceeds the scope of the chapter which sets out to discuss science versus religion.

from before "the psychological revolution" (Madsen 2011). A particular problem is to convince professional psychologists that they also produce, form and are ultimately responsible for the culture they are a part of. This is in many ways the answer to the proposed question: what is to be gained from a closer allocation between religion and psychology? The answer is possibly a closer recognition and feeling of responsibility of modern psychology as a type of religion or culture, which does not lean on the old distinction between large-P psychology and small-p psychology (Pickren & Rutherford 2010), but acknowledges the whole scope of psychology as something culturally embedded, like religion. Thus, psychology as a science and psychology as a religion, instead of psychology exclusively as science, might serve as a more solid foundation for professional ethical responsibility among psychologists. Of course, I am fully aware that it is highly unlikely that professional organizations like the American Psychological Association or the Norwegian Psychological Association would embrace this definition as they would understandably fear a loss of status and a setback in the everlasting encounter with professions like medicine over numerous authorizations. I do however think that, at the very least, the perspective of psychology as religion could and should be implemented, for instance, in the education of tomorrow's psychologists in order to raise their ethical awareness and understanding of what psychology historically and currently is.

Conclusion

In this chapter I have maintained that whether we choose to consider psychology as science or as religion depends on the historical perspective applied. Psychology's official history lesson predominately takes Wundt's experimental laboratory in Leipzig founded in 1879 as its point of departure. The impression one is left with today is thus that psychology belongs with the natural sciences. A road less traveled is to consider modern-day psychology as rooted in the strong cultural currents at the heart of the Reformation movement in sixteenth-century medieval Europe. The historical view of psychology as a science or related to religion determines how we look upon psychology today. For instance, mental health disorders would perhaps have been interpreted differently than "natural kinds" like our genes and neurochemical balances in the brain, and more rooted in the cultural sphere of meaning. Would this view again lead to a wholly different approach to the understanding and treatment of mental health problems? I can only pose these questions here as food for thought. The problem is that up until now, critical psychology[19] has struggled to get mainstream psychology to acknowledge psychology as part of the problem, as deeply culturally

[19] Critical psychology is a subdiscipline of psychology that "believes that mainstream psychology has institutionalized a narrow view of the field's ethical mandate to promote human welfare" (Fox, Prilleltensky & Austin 2009: 3).

embedded in the same social forces as the many ills it is trusted to treat and cure. A view of psychology as a religion as much as a science could help solve this deadlock by giving fewer reasons for operating with a clear distinction between psychology and culture and the individual and society. As the pioneer in *Völkerpsychologie* Moritz Lazarus so strikingly put it exactly 150 years ago: "We cannot emphasize the following enough, society does not consist of individuals as such, rather it is within and from society that individuals exist" (Lazarus in Diriwächter 2004: 90).

Bibliography

Bellah, R.N. 2008. *Habits of the heart: individualism and commitment in American life: with a new preface*. Berkeley, CA: University of California press.

Boring, E.G. 1957 [1950]. *A History of Experimental Psychology*. 2nd ed. New York: Appleton-Century-Crofts.

Bregman, L. 2001. The death awareness movement: Psychology as religion? In D. Jonte-Pace & W.B. Parsons (eds.). *Religion and Psychology: Mapping the Terrain*. London: Routledge. 319-332.

Brinkmann, S. 2008. Identity as Self-Interpretation. In *Theory & Psychology* 18 (3): 404-422.

Carrette, J. 2001. Post-structuralism and psychology of religion: The challenge of critical psychology. In D. Jonte-Pace & W.B. Parsons (eds.). *Religion and Psychology: Mapping the Terrain*. London: Routledge. 110-126.

Casanova, J. 2011. The Secular, Secularizations, Secularisms. In C. Calhoun, M. Juergensmeyer & J. VanAntwerpen (eds.). *Rethinking Secularism*. Oxford: Oxford University Press. 54-74.

Cruikshank, B. 1996. Revolutions within: self-government and self-esteem. In A. Barry, T. Osborne & N. Rose (eds.). *Foucault and political reason. Liberalism, neo-liberalism and rationalities of government*. Chicago: The University of Chicago Press. 231-251.

Danziger, K. 1990. *Constructing the Subject: Historical Origins of Psychological Research*. Cambridge: Cambridge University Press.

Dawkins, R. 1997. Is Science a Religion? Retrieved from http://thehumanist.org/humanist/articles/dawkins.html.

Diriwächter, R. 2004. Völkerpsychologie: The Synthesis that Never was. In *Culture & Psychology* 10 (1): 85-109.

Durkheim, É. 1971 [1912]. *The elementary forms of the religious life*. London: Allen & Unwin.

Ebbinghaus, H. 1908. *Psychology: An elementary textbook*. Boston: Heath.

Einstein, A. 1949. Religion and science. In A. Einstein (ed.). *The World as I See It*. New York: Philosophical Library. 24-28.

Foucault, M. 1978. Govermentality. In G. Burchell, C. Gordon & P. Miller (eds.). *The Foucault Effect*. London: Harvester Wheastheaf. 87-104.

Fox, D., I. Prilleltensky & S. Austin. 2009. *Critical Psychology: An Introduction*. 2nd ed. London: SAGE.

Freeman, M. 2001. Religion's Threat to Theoretical Understanding. In *Theory & Psychology* 11: 578-584.

Illouz, E. 2008. *Saving the modern soul: therapy, emotions, and the culture of self-help*. Berkeley, CA: University of California Press.

James, W. 1982 [1902]. *The Varieties of Religious Experience: A Study in Human Nature*. New York: Penguin.

James, W. 1985 [1892]. *Psychology: The Briefer Course*. Notre Dame, IN: University of Notre Dame Press.

Kvale, S. 1992. Postmodern Psychology: A Contradiction in Terms? In S. Kvale. (ed.). *Psychology and Postmodernism*. London: Sage. 31-57.

Kvale, S. 2003. The Church, the Factory and the Market: Scenarios for Psychology in a Postmodern Age. In *Theory & Psychology* 13 (5): 579-603.

Loss, C. P. 2002. Religion and the Therapeutic Ethos in Twentieth-Century American History. In *American Studies International* 40 (3): 61-76.

Madsen, O.J. 2011. *The Unfolding of the Therapeutic. The Cultural Influence of Psychology in Contemporary Society*. PhD. University of Bergen.

Miller, P. & N. Rose. 2009. *Governing the Present*. Cambridge: Polity Press.

Nafstad, H.E. 2008. Assumptions and values about human suffering in research and practice: Towards an area ethics of psychology. In S. Robinson & J. Strain (eds.). *Ethics For Living and Working*. Leister: Troubador Publishing. 52-63.

Parker, I. 2007. *Revolution in Psychology. Alienation to Emancipation*. Pluto Press: London.

Pawlik, K. & M.R. Rosenzweig. 2000. Psychological science: Content, methodology, history and profession. In K. Pawlik & M.R. Rosenzweig. (eds.). *International handbook of psychology*. Thousand Oaks, CA: SAGE. 3-19.

Pickren, W. & A. Rutherford. 2010. *A History of Modern Psychology in Context*. Hoboken, NJ: John Wiley.

Prilleltensky, I. 1997. Values, Assumptions, and Practices. Assessing the Moral Implications of Psychological Discourse and Action. In *American Psychologist* 52 (5): 517-535.

Rieff, P. 1987. *The triumph of the therapeutic: Uses of faith after Freud (with a new preface)*. Chicago, IL: University of Chicago Press.

Rieff, P. 2006. *Sacred Order/ Social Order: My Life among the Deathworks. Illustrations of the Aesthetics of Authority*. Charlottesville: University of Virginia Press.

Rose, N. 1996. *Inventing our selves: Psychology, power and personhood*. Cambridge: Cambridge University Press.

Smith, C. & M.L. Denton. 2005. *Soul searching: the religious and spiritual lives of American teenagers*. New York: Oxford University Press.

Taylor, C. 1992. *The ethics of authenticity*. Cambridge, MA: Harvard University Press.

Teo, T. 2005. *The critique of psychology: from Kant to postcolonial theory*. New York: Springer.

Tillich, P. 1957. *Dynamics Of Faith*. New York: Harper & Brothers Publishers.

Vitz, P.C. 1991 [1977]. *Psychology as religion: the cult of self-worship*. Grand Rapids, MI: W.B. Eerdmans.

Vitz, P.C. 1995. The Problem with Self-Esteem, *Catholic Education Research Center*. Retrieved from http://www.catholiceducation.org/articles/education/ed0001.html.

Weber, M. 1970. The Social Psychology of the World Religion. In H.H. Gerth & C.W. Mills (eds.). *From Max Weber: Essays in sociology*. London: Routledge & Kegan Paul Ltd. 267-301.

Wright, K. 2011. *The Rise of The Therapeutic Society*. Washington, DC: New Academia Publishing.

Chapter 7: Science without God

Ragnar Fjelland

Introduction: science vs. religion

From the very birth of modern science in the seventeenth century there has been tension between science and religion. This tension is symbolized by the trial of one of the founding fathers of modern science, Galileo Galilei. His defense of Copernicus's heliocentric system was condemned by the Catholic Church as heresy. He was forced to withdraw his assertions that the sun is the center of the universe and does not move, and that the earth is not the center of the universe and moves.

In 1609 Galileo had heard about a Dutchman who had constructed a "spyglass" by inserting two lenses into a tube. Based on this description he constructed his own telescope and directed it towards the sky to make observations of the stars and the planets. In 1610 he observed Jupiter for a longer period, and noticed that four objects moved around the planet. He concluded that Jupiter actually had four moons. His observations were published in *The Starry Messenger* the same year, and caused a sensation throughout Europe. Galileo regarded his observations as strong support for the heliocentric system. According to the orthodox view, based on the Scripture and the works of Aristotle, the universe could only have one center of rotation. That center was the earth, and all the heavenly bodies rotated around the earth as center.

The clash between Galileo and the church is the topic of Berthold Brecht's play *Life of Galileo*. In Brecht's play there is a scene where Galileo has a discussion with a philosopher and a mathematician. Galileo invites them to look through his telescope. However, the philosopher first wants to discuss if such objects can exist, and the mathematician adds: "You are aware, of course, that in the view of the ancients no star can revolve around any center other than the earth and that there can be no stars without firm support in the sky." A little later the mathematician adds: "One might be tempted to reply that if your tube shows something that cannot exist it must be a rather unreliable tube" (Brecht 2007: 22). In the end, they decline the invitation to look into the telescope.

The church no doubt saw that the heliocentric system undermined man's special status in the order of the universe. Brecht lets "the very old cardinal" express these worries in the following way:

> I'm not some nondescript being on some little star that briefly circles around somewhere. I walk with assurance on a firm earth, it stands still, it is the center of the universe, I am in the center, and the Creator's eye rests on me, on me alone. Around me, fixed to eight crystal spheres, revolve the fixed stars and the mighty sun, which was created to illumine my surroundings. And myself as well, in order that God may see me. Hence obviously and irrefutably, everything depends on me, man, the supreme work of God, the creature in the center, the image of God, imperishable (Brecht 2007: 32).

As early as 1616, the theologists who were consulted to investigate Copernicus' and Galileo's assertion that the sun is at rest at the center of the universe argued that this assertion was scientifically wrong, but the main error was that it explicitly contradicts many places in the Holy Scripture (McMullin 1998). For example, in Psalm 104 we can read about the immobility of the earth: "He set the earth on its foundations; it can never be moved" and in Ecclesiastes 1:5 we can read about the motion of the sun: "The sun rises and the sun sets, and hurries back to where it rises".[20]

From the point of view of religion the challenge from modern science is that some basic scientific results both contradict the Holy Scripture and deprive man of his special place in the cosmic order. The biological theory of evolution is an even more serious challenge than the heliocentric system. The Bible tells us that the universe was created in seven days, and that man is not an animal, but was created in the image of God. Therefore, the theory of evolution both contradicts the Scripture and reduces man to one species among numerous other species of animals.

Brecht's play points to one important difference between science and religion. Science is based on observation, and therefore scientific knowledge is partial and temporary. This is of course a limitation, but science develops and makes progress. Religion, on the other hand, is all-embracing and eternal. Although the Bible, for example, has a history, after all the texts were selected by human beings, the content is regarded as eternal and not subject to revision. In addition religion gives a total view of the universe, including man's place in the system, and it prescribes rules for the right conduct.

Therefore, the church at the time of Galileo did not only regard itself as an authority on religious questions, but of scientific questions as well. In principle, the Church might have given up a literal reading of the Bible, and admitted that some passages should be read symbolically. In particular, the Scripture must not be taken literally

[20] Both translations retrieved from http://www.biblegateway.com/.

when it comes to scientific questions. It is neither a textbook in cosmology nor in biology. This was probably not an option at the time of Galileo, and he was prosecuted for heresy.

However, if Galileo on his part had emphasized the restricted and temporary status of scientific knowledge, his confrontation with the church might have been avoided. This is what Copernicus had done one hundred years before. In his preface to *On the Revolutions of the Heavenly Spheres* he emphasized that the heliocentric system did not claim to be true, but was a hypothesis with the sole purpose of simplifying the astronomical calculations:

> For it is the job of the astronomer to use painstaking and skilled observation in gathering together the history of the celestial movements, and then–since he cannot by any line of reasoning reach the true causes of these movements–to think up or construct whatever causes or hypotheses he pleases such that, by the assumption of these causes, those same movements can be calculated from the principles of geometry for the past and for the future too [...] [F]or it is not necessary that these hypotheses should be true, or even probably; but it is enough if they provide a calculus which fits the observations (Copernicus 1990: 505).

Today it is generally acknowledged that the foreword was not written by Copernicus himself, but by his friend Andrew Osiander, who saw the book through the press. Probably Osiander's main motivation was to avoid prosecution by the Church.

Contrary to Copernicus (or Osiander), Galileo did not admit any uncertainty or limitations in his scientific knowledge. He rather insisted on the truth and certainty of his claims (McMullin 1998: 272). This is sometimes attributed to Galileo's stubbornness. It is probably true that Galileo was an obstinate person, but his position cannot be reduced to psychology. It rather reveals an important aspect of modern science that is often neglected. This aspect is the pursuit of *certainty*, and it applies in particular to the "higher sciences" (geometry, astronomy, mechanics) in contrast to the "lower sciences" (geology, medicine). The first category holds the mathematical sciences, whereas the latter comprises the non-mathematical (natural) sciences.

Nevertheless, Galileo himself did not regard his own science as opposed to religion. But instead of a literal reading of the Bible he regarded mathematics as the key to God (Redondi 1998: 201). Today we find an abundance of literature that uses science to support religion. For example, several physicists use modern physics and cosmology to show that the account of the creation of the universe in the Bible was roughly right. When we reinterpret the time perspective – they argue – it is basically compatible with the theory of the Big Bang. A prominent example is *The Science of God. The Convergence of Scientific and Biblical Wisdom*, written by the physicist Gerald Shroeder (1998). On the other hand we find biologists who argue that religious claims are incompatible with science, and that biology can even account for religion as a biological phenomenon.

In this chapter I shall restrict myself to dealing with the second kind of literature: biologists who argue that religion is incompatible with science. I shall deal with two evolutionary biologists, Edward Wilson and Richard Dawkins, and a molecular biologist, Jacques Monod. All three of them, including Monod, argue that science is not only incompatible with religion, but that evolutionary biology can explain religion, or, if one wishes, explain religion away. I will show that their arguments are not based on science, but on what I will call "scientific fundamentalism". Scientific fundamentalism may not be as evil as religious fundamentalism. However, the two positions have more in common than Wilson, Dawkins and Monod are aware of. To show this I will first deal with the origin of scientific fundamentalism in Galileo and Descartes. They knew that scientific fundamentalism actually represents "God's Eye View". Therefore, a proof of God's existence was imperative to Descartes's project. But, ironically enough, it is actually imperative to Wilson's, Dawkins' and Monod's projects as well. I will finally take a closer look at the consequences of a "science without God", and discuss if there is any place for religion.

Scientific fundamentalism

Although Galileo's observations of Jupiter's moons were important, they were not his most important contribution to science and philosophy. Observation is important in science, but it is not the distinguishing mark of modern science. For example, Aristotle was a skilled observer, whereas Galileo on several occasions emphasized that we should not rely too much on observations, because our senses may deceive us. Although it may look like a paradox, it is easy to see why Galileo had to stress the unreliability of our senses: if the heliocentric hypothesis is true, the earth moves through space with a tremendous speed. How can it be that we do not observe this motion? Galileo's answer was that we do not observe absolute, but relative motion (Galilei 1970). This is the *principle of relativity*. But who could believe this? Galileo actually tried to persuade his contemporaries not only to abstract, but to put the *abstract* world of mathematics in the place of the world that we live in and observe immediately. If we should characterize modern science by one sentence it would be this: to prefer the abstract world of mathematics to the concrete world that we observe.[21] This is what I have called "scientific fundamentalism".

This can be illustrated by using Galileo's metaphor of the book of nature. He emphasized that to read the book, we first have to learn the language in which it has been written. About this language he said:

> Philosophy is written in this grand book, the universe, which stands continually open to our gaze. But the book cannot be understood unless one first learns to comprehend the language and read the letters in which it is composed. It is written in the language of mathematics, and its characters are triangles, circles, and other geometric figures without which it is humanly impossible to understand a single word of it; without these, one wanders about in a dark labyrinth (Galilei 1970: 237-8).

This quotation is from his *Dialogue on two Chief World Systems*, published in 1630, where he defended the heliocentric hypothesis. However, he had almost ten years earlier introduced the division between primary and secondary qualities. Primary qualities are the qualities that can be described mathematically, like position and velocity, whereas the secondary qualities are the qualities we perceive, like colors, smell and taste. The choice of terminology was not accidental. On the contrary, "primary" and "secondary" indicate the ontological status of the denoted objects. In brief, only the primary qualities exist objectively, and may be the object of scientific investigations (Galilei 1957: 274).

Galileo's view was adopted by his younger contemporary René Descartes, who made it the very foundation of modern philosophy. In a famous passage from *Meditations* he contemplates a piece of wax. The wax has smell, color, shape, size, and it is hard, cold and tangible. But if it is placed near a flame, neither smell nor color remains. In fact, Descartes concludes that nothing of all the things that we perceive by means of our senses, remains. Therefore, these properties cannot belong to the wax itself. The only properties that may be attributed to the wax are that it is extended, flexible and malleable (Descartes 1971: 108). In Descartes's own terminology, that which exists objectively, is *res extensa*.

[21] Perhaps the first to point out this was the historian of science Alexandre Koyré (Koyré 1978). He influenced Edmund Husserl, who called it "Verhängnisvolle Missverständnisse als Folgen der Unklarheit über den Sinn der Mathematisierung" (Husserl 1962: 54) ("Portentous misunderstandings resulting from lack of clarity about the meaning of mathematization", Husserl 1970: 53). It is the main point in Paul Feyerabend's posthumously published book *Conquest of Abundance* (1999).

Man has a body, and therefore he is a *res extensa*, but in addition he has consciousness, and is a *res cogitans*. The *cogito* was the starting point of Descartes's philosophy, the self-evident point of departure of his whole philosophical system. But the cogito is not subject to the laws of the material world. Material bodies are governed by the laws of mechanics, whereas the cogito is governed by reason. As material beings we are subject to the laws of mechanics, but as conscious beings we are rational and free.

The Cartesian division of reality into one material and one immaterial part may serve as a foundation for a coexistence between science and religion: they simply belong to two different spheres of reality. In modern thinking God is therefore not directly present in the material world. Today we simply take this for granted, and if we meet cultures where it is not taken for granted, it looks rather odd to us. We find a good example of pre-modern thinking in Karen Blixen's *Out of Africa*. Her friend Denys had an airplane, which they used for pleasure. One day they had been out flying when an old Kikuyu, belonging to a local tribe, came up and started a conversation. He told them that they must have been very high, because he could not see them, but had heard the plane sing like a bee. When they confirmed that they had been very high, he asked if they had seen God up there. When they answered that they had not, he asked if they thought that they would be able to fly high enough to see God. When they expressed their doubts, the Kikuyu finished: "Then I do not know at all why you two go on flying" (Dinesen 2002: 245). Modern man knows better. He does not look for God in the material world.

Can science explain religion?

If the heliocentric system deprived man of a privileged place in the universe, Darwin's theory of evolution by natural selection was an even more serious blow to man's privileged position. It directly contradicts Genesis, according to which the universe was created in seven days, and man was created in the image of God. Even if evolution is something new relative to Galileo's and Descartes's materialism, the theory of evolution may also be regarded as materialistic. It explains the higher forms of life as having developed from lower forms without any plan, only through the mechanisms of variation and natural selection. In the end, all life has evolved from inanimate matter.

In later years evolutionary theory has also been used to explain animal behavior as a product of natural selection, known as "sociobiology". It has been extended to include human behavior, and is then known as "human sociobiology" (today also known as "evolutionary psychology"). The program is well represented in the following quotation from Edward Wilson's best-selling book *On Human Nature* (that won him the Pulitzer Prize), where he called it "scientific materialism":

> The core of scientific materialism is the evolutionary epic. Let me repeat its minimum claims: that the laws of the physical sciences are consistent with those of the biological and social sciences and can be linked in chains of causal explanation; that life and mind have a physical basis; that the world as we know it has evolved from earlier worlds obedient to the same laws; and that the visible universe today is everywhere subject to these materialist explanations (Wilson 1979: 208).

According to Wilson, human phenomena are subject to these materialistic explanations. For example, the human mind "[...] is a device for survival and reproduction, and reason is only one of its various techniques" (*ibid.* 3). Therefore, it does not belong to a different, immaterial sphere. It should be investigated by science.

This applies to all human phenomena, including ethics and religion. According to Wilson evolutionary theory can even explain the phenomenon of religion: "Most importantly, we have come to the crucial stage in the history of biology when religion itself is subject to the explanations of the natural sciences" (*ibid.* 200).

It goes without saying that according to Wilson's scientific materialism, religion is a delusion. Therefore, he must give a biological explanation of the phenomenon. However, one might think that evolutionary theory will have problems accounting for religion. A delusion will normally decrease the ability to survive and reproduce. For example, if a person is victim of the delusion that just praying will be sufficient to supply the food and shelter required to survive, he will probably not survive. But according to Wilson religion has a social function. It even solves one of the major problems in the Darwinian theory of evolution by natural selection. The problem is to account for the phenomenon of altruism. The properties that increase the individual's ability to survive and reproduce are favored by natural selection. A tendency to act altruistically towards other individuals will decrease this ability. A simple example will show this: an owl approaches a flock of sparrows. If a sparrow observes the owl first and warns his fellow sparrows, he will attract the attention of the owl and increase his risk of being the owl's prey. If the sparrow keeps silent, the chances of surviving will increase. Therefore, selfishness pays for the individual, and if this behavior has a genetic foundation, it is passed on to the offspring. The same applies to humans. In Wilson's own words, "[...] fallen heroes do not have children" (*ibid.* 159).

Although Wilson emphasizes the virtue of egoism, he admits that society needs a certain amount of altruism to work. Members of society must be willing to give up some of their self-interest in favor of the common good. Therefore, although religion is a delusion, it has the positive effect that it increases the solidarity of the group. Therefore, "[...] the highest form of religious practice, when examined more closely, can be seen to confer biological advantage" (*ibid.* 196).[22]

According to Wilson, the positive effect of religion will make it difficult to replace it by science. One may ask why it should be replaced if it has positive effects. Wilson does not ask the question. However, I think that he takes it for granted that there is no way back. When we have adopted "scientific materialism" and discovered that religion is a delusion, the "magic" does not work anymore. It is difficult first to tell people that God is dead, and later point to the positive social effects of believing in God. Therefore, the rational way is to put science in the place of religion. This applies to ethics as well, and at this point Wilson is explicit: "Above all, for our own physical well-being if nothing else, ethical philosophy must not be left in the hands of the merely wise" (*ibid.* 7).

Can science replace religion?

The view that science can – and ought to – replace religion was actually advocated before Wilson by the French molecular biologist Jacques Monod in his book *Chance and Necessity*. Monod won a Nobel prize in medicine for his important contributions to genetics, and in the book he discusses the philosophical questions related to molecular biology. Like Wilson and Dawkins he offers an evolutionary explanation of religion. His starting point is that man could not survive outside a tribal group. Even if he could, belonging to a tribal group gave a tremendous selective advantage. But this required rules to govern the behavior of the members of the group, and these rules required a mythical explanation to gain legitimacy:

> we are the descendants of these men, and it is probably from them that we have inherited the need for an explanation, the profound disquiet which forces us to search for the meaning of existence. That same disquiet has created all myths, all religions, all philosophies and science itself (Monod 1971: 167).

According to Monod myths and religion, and philosophical systems, are the price we have to pay for living as social animals. Although these are mostly cultural

[22] The selection of traits that are beneficial to the group, and not to the individuals, is called group selection. Many evolutionary biologists have argued that group selection is incompatible with Darwinism. According to their view, Darwinism only permits two kinds of altruism: kin altruism and reciprocal altruism. Kin altruism is to act altruistically towards close relatives, who share many of the genes, and reciprocal altruism is following the rule: "I scratch your back, and you scratch mine". Reciprocal altruism is not really altruism; it is more enlightened egoism: I do someone a favor if I am relatively certain that it will be paid back. The problem is that religion does not fit into any of these categories. Therefore, Richard Dawkins gives an alternative evolutionary explanation in *The God Delusion*.

constructs, they would not have preserved their stability if they did not have a biological foundation. Therefore, man has "an innate need for complete explanation whose absence causes a deep inner anxiety". This is the basis of the "old covenant", which is an animist tradition.

However, science breaks with this tradition. The problem is that although society is based on modern science and technology, we have not accepted its fundamental message. Science is based on the postulate of objectivity, but it has only won its place in society, "[...] in men's practice, but not in their hearts". It is now time to acknowledge the message of science, and form a new covenant based on the *postulate of objectivity*. Monod explains it this way:

> If he accepts this message in its full significance, man must at last wake out of his millenary dream and discover his total solitude, his fundamental isolation. He must realise that, like a gypsy, he lives on the boundary of an alien world; a world that is deaf to his music, and as indifferent to his hopes as it is to his suffering or his crimes (Monod 1971: 172-3).

The postulate of objectivity is actually based on an "ethical postulate". For some reason Monod does not formulate this ethical postulate. However, it looks as if the postulate is simply that knowledge and ethics should be kept strictly apart. This is actually implicit in the postulate of objectivity. We can only have knowledge of the world, and this knowledge does not tell us anything about what is right and wrong. This is actually Hume's old maxim that "is" does not imply "ought".

Science with God, and science without God

I agree with Monod in many respects. We are alone in the universe, and cannot rely on any external authority to tell us what is right and wrong. However, I shall argue that Monod is not radical enough. Although he acknowledges that his objectivity postulate is an ethical choice, it looks as if there are only two possibilities: either the "old covenant" or the postulate of objectivity. He does not say much about objectivity, but according to his description it is clear that by objectivity he means the same as Galileo and Descartes: the world as it exists independently of any human being. What is real is what complies with Galileo's primary qualities and Descartes's *res extensa*.

The world of science, the reality that Monod offers us (along with Wilson and Dawkins), is therefore a rather impoverished reality. In *Conquest of Abundance* Paul Feyerabend quotes the passage from Monod that I have quoted above, and adds: "The destruction caused by the progress of science cannot be described more clearly" (Feyerabend 1999: 6). To demonstrate that this is problematic he uses the rainbow as an example. According to Feyerabend the rainbow is real. It can be observed by independent observers, it can be painted and it can be photographed. However, it cannot be touched like a stone or a table. Feyerabend's point is that there are many things in the world that are real in the sense that they occur, are noticed, and have effects: dreams, stones, sunrises, rainbows, fleas, murders, errors (*ibid.* 7). These things have different properties and different consequences in different circumstances. Therefore, he gives the following advice:

> Grand subdivisions such as the subdivision real/unreal are thus much too simplistic to capture the complexities of our world. There are many different types of events, and "reality" is best attributed to an event together with a type, not absolutely (Feyerabend 1999: 8).

This should rather be our starting point: there are many different things, and many different perspectives. To admit only one perspective, the perspective of science, is what I have called "scientific fundamentalism".

But is the allegation of scientific fundamentalism really justified? Dawkins addresses the question in *The God Delusion*. He admits that in a certain sense scientists are fundamentalists. Similarly to Monod, he argues that "a scientist's belief in evidence is in itself a matter of fundamentalist faith" (*ibid.* 282). But according to Dawkins, scientists are no more fundamentalists than anybody else, because "fundamentalism" in this context is simply to require evidence when asserting that something is true. He uses an example to illustrate this: there is no difference between saying that New Zealand is in the southern hemisphere and saying that evolution is true. In both cases we will ask for evidence. But let us take a closer look at the two cases. In the case of New Zealand we do not need much theory to decide that the assertion is true. The example complies with a simple version of the correspondence theory of truth. It is worth keeping in mind, though, that it is true in a world inhabited and structured by human beings. It would not make sense to speak about New Zealand independently of human beings, and the globe does not come naturally divided into a northern and a southern hemisphere. They are not "natural kinds". To put it simply, when we say that it is true that New Zealand is located in the southern hemisphere, we already presuppose ordinary language in a world inhabited by human beings.

What about evolution, then? If we say that the theory of evolution is true, what does it correspond to? With some imagination we may come up with examples that illustrate evolution, for example successive stages of monkeys becoming more and more like human beings. But these are only scattered examples, and cannot be said to correspond to the assertion that life is the product of evolution. Whatever it corresponds to, it is different from the location of New Zealand in the southern hemisphere. If we go further, to modern theories of evolution, we move further away from the New Zealand example. What does Darwin's theory of evolution by natural selection correspond to? And what about Neo-Darwinism, the synthesis between Darwinian evolution and Mendelian genetics? Whatever the truth of these theories entails, we are far away from the example of New Zealand.

Dawkins's example conceals the fact that "the world of science" is different from our everyday world. Monod saw what Dawkins does not see, that the two "worlds" are fundamentally different. Monod rightly pointed out that it was the liberation from the everyday world that created modern science. There is nothing wrong with this. But it allows two errors. The first is not to see that they are different. The second is to take the world of science to be absolute. That is what "scientific fundamentalism" does.

Conclusion: we can do without religion

But Monod also left something out. Galileo and Descartes knew that to make "the world of science" absolute, we need God to establish the connection to our everyday world. The key is mathematics. Galileo's primary qualities and Descartes's *res extensa* have mathematical properties only. Both Galileo and Descartes used God as a guarantee for the truth of mathematics. God has written "the book of nature" in the language of mathematics.

There is an important similarity between fundamentalism in science and religion. God has revealed his thoughts in two books: the Bible and "the book of nature". However, there is only one correct way of reading both books. If we read them correctly, we will discover the truth. From a fundamentalist point of view neither the Bible nor "the book of nature" allows different interpretations.

It may look ironic that fundamentalist scientists are the most passionate opponents of fundamentalist religion (and Dawkins is one example). But it should be no surprise. We find the same between religions. Fundamentalist Christians are the most hostile critics of fundamentalist Islam, and vice versa. They are not opposed to their literal reading of the Koran. They only want to replace it by a literal reading of the Bible.

The metaphor of "the book of nature" is so powerful that Einstein used it even though he was an atheist. Once he compared the universe to a large collection of books. When we understand the language and can read the books, we understand the universe. It is expressed in his famous quotation: "Subtle is the Lord, but he is not malicious" (Pais 1982). We find a similar view in Stephen Hawking. In *A Brief History of Time* he describes the ambitions of a "Theory of Everything", and concludes that when these ambitions have been realized, "Then we know the mind of God" (Hawking 1986).

If we give up scientific fundamentalism, science cannot be used to show that there is no place for religion. However, the fact that science does not exclude religion does not represent a positive argument in favor of religion. We are left with at least three possibilities:
1. we adopt a fundamentalist religion,
2. we adopt a non-fundamentalist religion, or
3. we do better without any kind of religion.

Concerning (1), it follows from what I have said previously that a commitment to a fundamentalist version of religion is out of the question. A literal reading of for example the Bible is incompatible with modern science. But there are also moral reasons for not subscribing to a literal reading of the Bible. I agree with Dawkins when he quotes the liberal Bishop John Shelby Spong who has said that they who base their ethics on a literal reading of the Bible have either not read it or not understood it (Dawkins 2006: 237). The God of the Old Testament is the warrior God, who orders mass murders and countless atrocities. One may argue that the Old Testament is only of historical interest. But then one leaves the literal reading of the Bible, and is hardly a fundamentalist. Nevertheless, the God of the Old Testament is the same as the God of New Testament. Dawkins points out, though, that Jesus, if he ever lived, was "a great ethical innovator" with his gospel of love (Dawkins 2006: 250). However, the problem with all fundamentalist religion today – be it Christian, Muslim or Jewish – is that they neglect the gospel of love, and rather appeal to hatred towards the infidels who do not worship their God. Therefore, there are both scientific and moral reasons to exclude fundamentalist religion.

What about (2) and (3)? It may be useful to go back to where I began, with the scientific revolution of the seventeenth century. This is usually taken to be the root of modernity. However, in *Cosmopolis: The Hidden Agenda of Modernity* (1990), Stephen Toulmin argues that this traditional view is problematic. He agrees that the birth of modern science, and of modernity, can be dated to the first part of the seventeenth century. However, he argues that modernity has two different roots. One is the scientific revolution of the seventeenth century, with people like Galileo, Descartes and Newton, while the other is the Renaissance humanism of the 15th century, with people like Erasmus and Montaigne. As the term "humanism" indicates, it was based on a human perspective, characterized by an awareness of the limits of one's own perspective, the acceptance of uncertainty and the imperfection of man, and, therefore, a tolerance towards other opinions. However, the scientific revolution was not a continuation of Renaissance humanism. According to Toulmin it was rather a counter-Renaissance. The human perspective was replaced by the ideal of an absolute perspective, which defined objectivity in modern science, as I have explained previously.

Although God was probably always present in the mind of fifteenth century man, we see that it is possible to describe many of the values without invoking God. They are based on a human perspective, and therefore, they are contrary to traditional religion, which is based on an external authority. But it has one thing in common with traditional religion: they only make sense when man is regarded as belonging to something larger than himself. However, instead of being subordinate to a supreme being, man should rather be regarded as part of a community of fellow human beings, and a part of nature. We may be alone in the universe, but we are not like homeless gypsies, as Monod alleged. The earth may be a tiny planet in the outskirts of the universe, but it is nevertheless the center of *our* universe, and therefore our home. Therefore, I will end this article with a quotation from Einstein that Dawkins uses as an introduction to his chapter on ethics:

> Strange is our situation here upon earth. Each of us comes for a short visit, not knowing why, yet sometimes seeming to divine a purpose. From the standpoint of daily life, however, there is one thing we do know: that man is here for the sake of other men – above all for those upon whose smile and well-being our own happiness depends, and also for the countless unknown souls with whose fate we are connected by a bond of sympathy (Einstein in Dawkins 2006: 209).

Bibliography

Brecht, B. 2007. *Life of Galileo*. Translated by W. Sauerlander & R. Manheirn. Retrieved from http://buehnenkunst.bu.ohost.de.

Copernicus, N. 1990 [1543]. *On the Revolutions of the Heavenly Spheres*. Chicago: Encyclopædia Britannica, 2nd edition.

Dawkins, R. 2006. *The God Delusion*. London: Bantam Press.

Descartes, R. 1971 [1641]. Meditations, in R. Descartes: *Discourse on Method and the Meditations*. Harmondsworth: Penguin Classics.

Dinesen, I. (Karen Blixen). 2002 [1937]. *Out of Africa*. New York: Random House.

Feyerabend, P. 1999. *Conquest of Abundance*. Chicago: University of Chicago Press.

Galilei, G. 1957 [1623]. The Assayer. In S. Drake (ed.). *Discoveries and Opinions of Galileo*. New York: Doubleday & Company.

Galilei, G. 1970 [1630]. *Dialogue Concerning the Two Chief World Systems*. Berkeley/Los Angeles: University of California Press.

Hawking, S. 1986. *A Brief History of Time*. London: Bantam Press.

Husserl, E. 1962. *Die Krisis der europäischen Wissenschaften und die tranzendentale Phänomenologie*. Haag: Martinus Nijhoff.

Husserl, E. 1970. *The Crisis of European Science and Transcendental Phenomenology*. Evanston: Northwestern University Press.

Koyré, A. 1978 [1939]. *Galileo Studies*. London: Harvester.

McMullin, E. 1998. Galileo on science and scripture. In P. Machamer (ed.). *The Cambridge Companion to Galileo*. Cambridge: Cambridge University Press. 271-347.

Monod, J. 1971. *Chance and Necessity. An essay on the Natural Philosophy of Modern Biology*. New York: Alfred Knopf.

Pais, A. 1982. *"Subtle is the Lord...": the Science and the Life of Einstein*. Oxford: Oxford University Press.

Redondi, P. 1998. From Galileo to Augustine. In P. Machamer (ed.). *The Cambridge Companion to Galileo*. Cambridge: Cambridge University Press. 175-210.

Schroeder, G.L. 1998. *The Science of God. The Convergence of Scientific and Biblical Wisdom*. New York: Broadway Books.

Toulmin, S. 1990. *Cosmopolis. The Hidden Agenda of Modernity*. New York: The Free Press.

Wilson, E. 1979. *On Human Nature*. New York: Bantam Books.

Chapter 8: Science and religion, natural and unnatural

Barbara Herrnstein Smith

Recent years have seen the publication of a number of books and articles offering specifically "cognitive" and "evolutionary" accounts of the origins of religious beliefs (see e.g. Atran 2002, Boyer 2001, Dennett 2006, Lawson & McCauley 1990). These accounts explain many widespread religious concepts (gods, an afterlife, divinely ordained moral commandments and so forth) as products or by-products of the automatic, unconscious operations of innate, universal mental (or "cognitive") mechanisms that evolved in humans under Stone Age conditions. While such approaches to the understanding of religion have considerable intellectual interest, questions can be raised about their key assumptions, claims and methods and also about how those pursuing and promoting them – anthropologists, philosophers, psychologists and others – unfold their broader social and intellectual implications. I have dealt elsewhere with an array of theoretical and methodological problems presented by cognitive-evolutionary explanations of religion, especially in view of what I see as richer, more empirically responsive pragmatist and constructivist understandings of human cognition as well as more broadly informed accounts of the various phenomena (practices and institutions as well as concepts) we associate with the term "religion".[23] Here I shall focus on the enlistment of those explanations in the service of sharp but dubious contrasts between religion and science.

A new "cognitive" contrast

In the final chapter of his book *Religion Explained*, anthropologist Pascal Boyer begins a discussion of the relation between religion and science by rejecting a simple opposition between the two. "It is by no means clear," he writes, "that there is such a thing as "religion" in the abstract." Rather, he continues, invoking the strenuously mentalist ("cognitivist") account detailed elsewhere in his book, "[t]here are many mental representations entertained by people, many acts of communication that make them more or less plausible, many inferences produced in many contexts." He goes on to make parallel observations of science:

> Science too is a cultural thing, that is, a domain of mental representations that happen to be entertained by a number of human minds. There is no science as such but rather a large set of people with particular activities, a particular

[23] See Smith 2010. The present essay draws on sections of this book.

> database that is stored in a particular literature, and a particular way of adding to or modifying that database (Boyer 2001: 320).

The religion-science symmetry soon dissolves, however, in favor of a familiar story of triumph and routing. In the West, Boyer writes, a "monopolistic doctrinal religion" (evidently Roman Catholicism)

> made the crucial mistake of meddling in empirical statements of fact […] In every instance where the Church has tried to offer its own description of what happens in the world *and* there was some scientific alternative on the very same topic, the latter has proved better. Every battle has been lost and conclusively so.

The apparent nominalism also dissolves in favor of a familiar generalized and essentialized concept of "science" and, with it, the assertion of its fundamental epistemic opposition to "religion," similarly generalized and essentialized: "Science showed not only that some stories about the formation of planets were decidedly below par but also that there was something dramatically flawed *in principle* about religion as a way of knowing things and that there was a better way of gathering reliable information about the world." Boyer concludes his discussion by reframing the religion-science relation as a contrast between the inherent attractiveness of religious concepts to the human mind and the unnaturalness of scientific thinking. Given our evolved mental dispositions, he writes, religious concepts are "a *likely* thing" for humans whereas scientific thought, being cognitively "unnatural," is an "*unlikely*" thing statistically and in fact quite rare among humans (Boyer 2001; see also Boyer 1994).

To support the terms of this dichotomy, Boyer cites an essay by philosopher Robert McCauley titled "The Naturalness of Religion and the Unnaturalness of Science," which itself echoes the arguments (and title) of a book by biologist Lewis Wolpert, *The Unnatural Nature of Science* (1993). These shared accounts and duplicated arguments are routinely cited by those promoting cognitive-evolutionary accounts of religion and recur in public debates over the relation between religion and science. The cogency of the natural-unnatural *topos* is, accordingly, worth some attention. I examine it here primarily in the version presented by Robert McCauley in the essay just mentioned but I also note its relation to more general contrasts between science and religion.

At the beginning of his essay, McCauley remarks that it is provoked by scholars in the field of religious studies who maintain that, because religion is not – or not simply – a natural phenomenon, its study requires methods other than those of the natural sciences. Seeking to turn the tables on such arguments, his intention, he writes, is to demonstrate that religion is, on the contrary, something supremely natural while it is actually science that is unnatural. As is often the case with polemical table-turnings, however, the reversal here does not come off altogether smoothly.

McCauley's demonstration, as he lays it out, consists of a series of strongly contrastive characterizations appealing to apparently straightforward observations supplemented by references to historical and experimental evidence. Thus he maintains that, from the fact that religion is found in all times and cultures, we may conclude that religious beliefs require nothing but the universals of human nature to spring up while, conversely, given the rarity of science, we may conclude that scientific thinking is essentially contrary to human nature. Or, he observes, inasmuch as science requires literacy, complex social arrangements, educated elites and technical means for preserving and transmitting knowledge, it is fundamentally "cultural" while, conversely, since religion requires nothing but basic cognitive abilities, it is "natural." Or again, the fact that religious concepts are easy to learn and remember and are quickly acquired even by young children indicates that such concepts conform to innate intuitions, while the fact that scientific concepts are hard to learn and take specialists years to master is evidence that they are counterintuitive and demand exceptional forms of cognitive discipline (McCauley 2000).[24]

Conceptual oversimplification and historical forgetfulness

These contrasts are in some ways plausible-sounding, draw on familiar observations and are presented by McCauley with a string of references to the psychological literature. The distinctions and alignments on which they are based, however, involve crucial conceptual oversimplification and historical obliteration. For one thing, it is not clear that comparable matters are being compared here, or that the comparisons are as even-handed as they could be. Thus, at the simplest level, we may ask what exactly it is in "religion" that children acquire so easily and in "science" that most people never come to master. To be sure, while many children can recite their

[24] Some of these arguments are significantly elaborated, modified and nuanced in McCauley 2011. The central contrastive claim remains, however, and some of the modifications evade important objections without overcoming them. A key modification is the replacement of the term "natural" in these formulations by a new concept, "*maturationally* natural," defined as an action requiring no instruction or artifacts, performed spontaneously and unconsciously, and found cross-culturally. The result is that what figured in the original essay as *evidence* for claiming that some religious idea is natural (for example, that it comes easily to children) is now part of the *definition* of naturalness. The modification avoids troublesome questions about how the naturalness of a cognitive process (or product) is being defined or could be established, but it turns many of McCauley's earlier arguments into empty tautologies. McCauley also remains unclear on the technically crucial question of whether or not "natural" (or now "maturationally natural") should be understood as equivalent to (or implying) "innate" in the sense of genetically specified.

prayers with ease and conviction, few could explain Einstein's Unified Field Theory. But many children who can recite the multiplication table at the drop of a hat would have considerable difficulty explaining the Doctrine of the Trinity. The fact, stretched here in the service of an exaggerated contrast, is that certain types of ideas and verbal routines – religious, scientific and other – are acquired readily while others (again, from any and all domains of thought) require a highly specialized education and relatively long apprenticeship for their mastery. The existence of such differences is not in question. What is dubious is the clear alignment of the first type with what are identified as specifically "religious" ideas and the second with what are identified as specifically "scientific" concepts.

Second and more fundamentally, in McCauley's essay as also in Boyer's and Wolpert's books, the sharp contrasts between a cognitively unnatural "science" and a cognitively natural "religion" require the usage of these terms in vague, shifting, overly broad, overly restrictive and otherwise tendentious ways. For example, while natural religious beliefs evidently include, for McCauley, everything from ancient sun-gods to the contents of parish catechisms, he insists on a historically, culturally and epistemically quite narrow understanding of science – which, of course, begs the question of the alleged primitiveness and ubiquity of the former and the alleged unnaturalness and rarity of the latter. Commenting on these objections in his recent book, *Why Religion is Natural and Science is Not*, McCauley writes that his (and Wolpert's) narrow definition of science as, in effect, the body of established Western scientific concepts and explanations is justified because "*that* is the science that most participants compare with (usually, modern Western) religion" (McCauley 2011: 89). Assuming that what McCauley means by "participants" here are scholars and laypersons participating in controversies over science and religion, then, to maintain due parity of reasoning, he should be identifying *religion* not with the naïve beliefs of children and other unsophisticated folk but with what participants in these controversies – including scholars of religion, theologians and educated churchgoers – typically mean by the term when they compare it in various ways to (usually, modern Western) *science* (see e.g. Hart 2009). If he did so, he would have to include some conceptually quite subtle and complex ideas – not to mention practices, institutions and intellectual and cultural elaborations, from Greek dramas and Gothic cathedrals to *Paradise Lost* and the sonnets of Gerard Manley Hopkins – that are no more common, spontaneous, easily produced or easily acquired than those generally associated with modern Western science.

In McCauley's essay as elsewhere in contemporary writings on these issues, sharp contrasts between science and religion also require the forgetting of quite a bit of recorded human history, notably the extensive historical overlaps and continuities between ideas, practices and institutions that are properly and reasonably included in the reference of each of the two terms. Among other things forgotten are the close intellectual as well as institutional ties between Western science and the Catholic Church for the better part of the past millennium. Historians of the subject remind us that much of what we now call science – pursued in the past as "natural philosophy" – was developed in medieval universities originally based in monastic orders and that recognizably scientific pursuits remained theologically oriented long afterward (see Harrison 2006, Olson 2004). As late as the eighteenth century, nature was studied systematically and empirically – by, among others, Isaac Newton – on the assumption that it embodied divine purpose and with the aim of revealing just how it did so. Historians also note that a number of familiar ideals and ideologies of modern Western science, such as the unity, progress and perfectibility of knowledge, are the fairly direct heritage of Christian doctrine, initially transmitted through the medieval universities and extended later by Enlightenment and evolutionary narratives of human rationality and development (see Noble 1992). These and other ideals and ideologies shared by Western science and monotheistic religions – including asceticism and patriarchy – seem to reflect more general human tendencies: for example, the inclination of people everywhere to construct teleological, meliorist narratives or to suppose that a strong male presence is required for important works of the mind or spirit.

Cognitive commonalities

A continuity of cognitive processes in the practices of religion, science and everyday life – along with observations on the generality of cognitive tendencies among humans – is both affirmed and denied by those proposing the natural-unnatural dichotomy that concerns us here. Thus, while Boyer emphasizes that religious persons are not essentially different from nonreligious ones in essential cognitive functions, he also maintains the exceptional cognitive and motivational character of scientists. The crucial point, he argues, is that, because of their special training, disciplined individual efforts and the unique normative system that defines their community, scientists come to act in ways that supersede their species-characteristic cognitive dispositions and impulses (Boyer 2001). That may be true. But the same could be said of, among others, Buddhist monks, classical scholars and Oxford-educated analytic philosophers, each of whom, given their special training, disciplined individual efforts and the distinctive normative systems that define their respective communities, could (and often do) make the same claims about their transcendence of ordinary human limits, cognitive and other. Scientists as a group may be unusual in these ways, but so also, it seems, are various other sets of people, in which case the claim of unnaturalness for scientific thinking would have to be extended more broadly. But if the cognitive activities of a good portion of humanity are species-transcendent, then species-transcendence would have to be counted as more or less natural to humans – something that, in fact, a good many humanistic thinkers have maintained.

Like Boyer, McCauley is equivocal on the question of the exceptional nature of scientific thinking, both acknowledging that scientists "exhibit the same cognitive biases and limitations that other human beings do" but also arguing that, unlike other human beings, they "get around" such biases and limitations. This, McCauley maintains, is because scientists have special "tools (such as literacy and mathematical description)" and because institutionally established norms encourage them to "seiz[e] opportunities to criticize and correct each other's work" (McCauley 2000: 66-67). The tools and norms that McCauley invokes are certainly significant in limiting the negative effects of scientists' cognitive liabilities. Their operation, however, is not as simple as he implies, nor their effectiveness as decisively differentiating. Among other epistemically significant cognitive tendencies that scientists share with religionists and humans more generally are animism, anthropomorphism, overgeneralization, essentialism, reification, hypertrophy, binary thinking, hierarchical thinking, linear-causal thinking, teleological thinking and a tendency to divide the social or intellectual world into communities of good/right *us* and bad/wrong *them*. One of the most significant of these shared cognitive tendencies is so-called confirmation bias, our tendency to notice and remember what confirms our established beliefs and to overlook or forget what contradicts them. Thus scientists may rationalize experimental anomalies and failed predictions in regard to current scientific theories in ways that resemble the belief-preserving rationalizations of religionists in regard to articles of faith and supposed biblical prophecies (see De Cruz & De Smedt 2007, Smith 2010). Also, while it is true that epistemically beneficial communal norms, such as accurate observation, precise statement or mutual criticism, are established and sustained among groups of scientists, it is also true that groups of scientists tend to share more intellectually dubious theoretical assumptions and related communal habits – or, in effect, biases – of perception and classification. The belief-preserving, dissonance-avoiding, innovation-discouraging, paradigm-hardening operations of these tendencies among scientists have been documented and theorized by historians and sociologists of science for close to a century, and most contemporary scholars of science reject the idea that scientific norms and methods effectively overcome – as opposed to check or limit – various types of human cognitive bias (see e.g. Fleck 1979, Henrich *et al.* 2010, Lightman & Gingerich 1992).

A dubious distinction

At a central point in his argument for the unnaturalness of scientific thinking, McCauley – closely following Lewis Wolpert here – insists that, to appreciate the "rarity" of science, we must not "confuse" it with "technology." "The crucial point," he writes, "is that the practical orientation of technology and the abstract theoretical interest in understanding nature that characterizes science are not the same aims [...] Science is finally concerned with understanding nature for its own sake and not merely for its effects on us" (McCauley 2000: 68, 71). But the admonition not to confuse science and technology, though familiar, is not so easy to heed. On the contrary, distinguishing them at all requires some significant retrospective tinkering. Most of the specialized pursuits we now associate with Western science, including anatomy, botany, chemistry and physics, developed in close conjunction with technical problem-solving in such perennial human activities as healing, agriculture, navigation and warfare. A tradition and image of gentlemen investigators seeking an understanding of the workings of nature "for its own sake" emerged in the seventeenth century, largely in the science academies of England and Europe, but the conjunction of epistemic pursuits with practical activities has continued unabated. Indeed, with the increased dominance of large-scale scientific ventures funded mainly by governmental, industrial and commercial agencies, any effort to mark off a realm of scientific thinking pursued independent of "a practical orientation toward technology" can only be arbitrary and artificial.[25]

The separation of science from technology required to sustain the unnatural/natural contrast with religion is not only conceptually strained and historically dubious but poses a considerable technical puzzle for the evolutionary-cognitive theorists promoting it. For, given the identification of science with a cognitively unnatural "abstract theoretical interest in understanding nature" "for its own sake," the question arises as to how, from an evolutionary perspective, such an enterprise – that is, one with no fitness-enhancing material advantages or connection to individual interests – could have arisen in the first place and why it has survived at all among humans. Indeed, McCauley seems to be led by just such considerations to represent science (as he defines it) as something quite fragile in competition with religion (as he defines it). He writes:

> In the global marketplace of ideas [...] some views have natural disadvantages. Science, with its esoteric interests, its counterintuitive claims, and its

[25] McCauley's defense of the sharp distinction he draws between science and technology (McCauley 2011) is equivocal: he acknowledges the extensive connections, overlaps and inextricabilities noted above but does not acknowledge (or does not recognize) their force for his and Wolpert's definitions of "science" and thereby also for their central claim concerning the "rarity" and "unnaturalness" of science relative to human cognitive processes and products.

> specialized forms of thinking, certainly seems to qualify. [Some scholars...] hold that science was once lost and had to be reinvented. One consequence of my view is that nothing about human nature would ever prevent its loss again (McCauley 2000: 82).[26]

This sounds rather ominous. There is, however, good reason to doubt that the survival of science, non-tendentiously defined, is as precarious as this suggests. Certainly the idea that the West might always return to the Dark Ages is exaggerated, forgetting the immense practical benefits, individual and communal, attached to existing scientific ideas, models and explanations and ignoring the continuously accelerated secularizing trends throughout the Western world. Indeed, a major movement in Western theologies is toward religion-science compatibilism (see e.g. Polkinghorne 2005), not a rejection of scientific knowledge, much less the extinction of scientific activity suggested by McCauley's grim vision.

Cognitively unnatural science?

Given the evolutionary dynamics of what McCauley and various other Darwinian theorists invoke as "human nature," the cognitive springs of science – even at its most esoteric and abstract – do not appear all that unnatural or even especially remote from the springs of religion, non-tendentiously defined. On the contrary, it seems clear that the array of distinctively – but also characteristically – human practices and techniques that we now call science arose in the course of efforts by our ancestors to solve practical problems of survival and that such practices and techniques (for example, abstract reasoning, measurement and standardized notational systems) were shaped cognitively, as well as culturally and materially, by their effectiveness in serving those ends. Rather than technology being, as McCauley and Wolpert typically represent it, a by-product of science, the reverse seems closer to the truth. That is, what they frame as the essence of science, "the abstract theoretical interest in understanding nature" "for its own sake," appears to be an offshoot of technology, a by-product of cognitive capacities and tendencies that evolved for more practically oriented activities.

Thus McCauley and Wolpert's "science" or "scientific thinking" is what Pascal Boyer calls, in regard to religious concepts and practices, "parasitic": that is (as Boyer explains it), a type of activity that emerges and persists among humans not because it confers any fitness benefits itself but because it recruits cognitive and other faculties or impulses that conferred such benefits in the course of human evolution.

[26] These observations appear, dramatically enough, as the closing sentences of McCauley 2011.

It appears, in other words, that, like other such activities (for example, performing music, playing chess or having sex for its own sake), the pursuit of pure knowledge for its own sake employs cognitive faculties and responds to bodily impulses that may have no fitness-related functions but the exercise and satisfaction of which are pleasurable in themselves. As a number of scientists have themselves observed, a scientist can derive deep pleasure just from his or her construction of a conceptually elegant, empirically confirmed explanatory model, quite apart from any practical applications that such a model might yield.

As it happens, developmental-cognitive psychologist Alison Gopnik has advanced a view of science as cognitively "parasitic" in just this way. "Science is successful," she writes, citing recent experimental findings, "because it capitalizes on a more basic human cognitive capacity," what she calls "the theory formation system drive." The fulfillment of that drive, Gopnik maintains, yields the deep satisfaction that humans, including young children, characteristically experience in the production of good explanations. She remarks: "Science is thus a kind of epiphenomenon of cognitive development. It is not that children are little scientists [a view that Gopnik advances elsewhere] but that scientists are big children," getting, in effect, a rush or a high from the fulfillment of an elementary drive; she compares it explicitly to sexual pleasure (Gopnik 2000: 300-301).

Much in Gopnik's account can be disputed. It is rather heavy on dubiously postulated drives and mental systems, and the orgasmic high that she claims is produced specifically by the generation of a good scientific or proto-scientific explanation is not clearly distinguishable from the satisfaction elicited by the successful completion of any strenuous intellectual (or creative or performative) venture or, indeed, from the successful execution of difficult physical (for example, athletic) feats. But her observations make clear that not all cognitive scientists are persuaded that scientific thinking is cognitively unnatural; and her account of scientific explanations suggests that what McCauley calls "the cognitive foundations" of pure science may be no more unnatural for humans than the cognitive foundations of writing poetry, playing the flute or solving cryptograms.

The differences between "science" and "religion," each duly historically defined and duly comprehensively indicated, remain both profound and important; and, of course, the political and intellectual stakes in distinguishing them appropriately are sometimes very high. But here as elsewhere the better way to go – better in both the political short run and the intellectual long run – is careful delineation and discrimination, not tendentious characterization, dubious dichotomy or exaggerated contrast.

Bibliography

Atran, S. 2002. *In Gods We Trust: The Evolutionary Landscape of Religion*. Oxford: Oxford University Press.

Boyer, P. 1994. *The Naturalness of Religious Ideas: A Cognitive Theory of Religion*. Berkeley, CA: University of California Press.

Boyer, P. 2001. *Religion Explained: The Evolutionary Origins of Religious Thought*. New York: Basic Books.

De Cruz, H. & J. De Smedt. 2007. The Role of Intuitive Ontologies in Scientific Understanding. In *Biology & Philosophy* 22: 351-368.

Dennett, D.C. 2006. *Breaking the Spell: Religion as a Natural Phenomenon*. New York: Viking.

Fleck, L. 1979 [1935]. Genesis and Development of a Scientific Fact. Edited by T.J. Trenn & R.K. Merton. Chicago: University of Chicago Press.

Gopnik, A. 2000. Explanation as Orgasm and the Drive for Causal Knowledge. In F.C. Keil & R.A. Wilson (eds.). *Explanation and Cognition*. Cambridge, MA: MIT Press, 2000. 299-324.

Harrison, P. 2006. "Science" and "Religion": Constructing the Boundaries. In *Journal of Religion* 86(1): 81-106.

Hart, D.B. 2009. *Atheist Delusions: The Christian Revolution and Its Fashionable Enemies*. New Haven, CT: Yale University Press.

Henrich, J. *et al.* 2010. The Weirdest People in the World? In *Behavioral and Brain Sciences*, 33: 61-83.

Lawson, E.T. & R.N. McCauley. 1990. *Rethinking Religion: Connecting Cognition and Culture*. Cambridge: Cambridge University Press.

Lightman, A. & O. Gingerich. 1992. When Do Anomalies Begin? In *Science*, 255(5045): 690-695.

McCauley, R.N. 2000. The Naturalness of Religion and the Unnaturalness of Science. In F.C. Keil & R.A. Wilson (eds.). *Explanation and Cognition*. Cambridge, MA: MIT Press. 61-85.

McCauley, R.N. 2011. *Why Religion is Natural and Science is Not*. New York: Oxford University Press.

Noble, D.F. 1992. *World without Women: The Christian Clerical Culture of Western Science*. New York: Knopf.

Olson, R. 2004. *Science and Religion, 1450-1900: From Copernicus to Darwin*. Baltimore, MD: Johns Hopkins University Press.

Polkinghorne, J. 2005. *Exploring Reality: The Intertwining of Science and Religion*. New Haven, CT: Yale University Press.

Smith, B.H. 2010. *Natural Reflections: Human Cognition at the Nexus of Science and Religion*. New Haven, CT: Yale University Press.

Wolpert, L. 1993. *The Unnatural Nature of Science*. Cambridge, MA: Harvard University Press.

Chapter 9: Immortality

An essay on science, technology and religion

Kjetil Rommetveit

> Age steals away all things, even the mind.
>
> *Virgil*

> The mission of the Immortality Institute is to conquer the blight of involuntary death.
>
> *Immortality Institute 2004*

A common article of faith in Western civilization has it that there exists, and should exist, a sharp line of demarcation between science and religion. The general progress of the sciences, especially the empirical and experimental, is associated with or even seen as a "cause of", secularization and a general decline in religiousness. This theme is itself a variation of a strong commitment once made in the West: when scientific knowledge increases, faith and superstition decreases. Such views can hardly be attributed to "science itself" or to "religion itself": many practicing scientists discover God in Nature, and many religious persons take a keen interest in science. Rather, the demarcation of science and religion is an outcome of long-lasting efforts and practices intrinsic to Western societies.[27] This chapter explores an idea and phenomenon in which spaces between science and religion collapse: immortality. Recent years have seen a renewed turn towards *engineered immortality* as a serious goal for research and innovation, to the extent that it is becoming a *leitmotif* for the 21st century. Huge amounts of public and private money have been invested (especially in the US), and researchers have tuned their experiments towards new goals: delayed senescence, anti-aging treatments, cryo-preservation, up-and-down-loading of consciousness to computers and digital networks. The most ardent promoter of engineered immortality, transhumanism,[28] is a mixed bunch of social visionaries, techno-prophets, practicing scientists, entrepreneurs,

[27] What Bruno Latour (1993) has called "work of purification".

[28] The transhumanist web-page humanity+ defines transhumanism as follows: (1) The intellectual and cultural movement that affirms the possibility and desirability of fundamentally improving the human condition through applied reason, especially by developing and making widely available technologies to eliminate aging and to greatly enhance human intellectual, physical, and psychological capacities. (2) The study of the ramifications, promises, and potential dangers of technologies that will enable us to overcome fundamental human limitations, and the related study of the ethical matters involved in developing and using such technologies (http://humanityplus.org/learn/transhumanist-faq/#answer_19).

businessmen and policy makers. Although not a conventional movement, it has gained a number of followers among groups such as scientists, engineers, artists, gamers, computer scientists, hackers and ethicists. The recent emergence of the transhumanist movement coincides with a major technoscientific program to make "converging technologies". Nano-, bio-, cogno-, info-, robo-: these are the prefixes in the naming of powerful, yet single-standing, new technologies. But imagine if they could all be united, could converge: solutions to our most pressing problems could be right around the corner, including old age, bad health, suffering and death.

This essay puts to the test the hypothesis that immortality, whereas clearly an esoteric and transcendent phenomenon, also possesses a number of earthly, social and cultural characteristics. I make three general presuppositions. First, that immortality aims to overcome a fundamental state of human existence, namely its finite character. Transcendence means to *rise above and beyond:* one's time, one's society or one's life. Second, that immortality is a *goal* for human achievement and excellence, in many ways *the* ultimate achievement. Third, that such transcendence relates in complex ways to the society in which it subsists, and to the cosmology of that society. The quest for immortality is a matter of personal ambition and longing, but at the same time an expression of shared values, goals and commitments. Transcendence, even where it ventures deep into Nature, explores the Universe or faces the Gods, must still be embedded in society. Even to the extreme, the drive for transcendence is bound up with social interests and structures of power.

That said; let me immediately add the following: immortality is a highly heterogeneous phenomenon and may take a great number of forms. This text deals with one subspecies that I shall term *socio-technical immortality*: the ways in which outstanding individuals inscribe themselves into the collective coordinates of ultimate reality through science and technology. The immortal individual must conquer reality and inscribe himself into his people's most basic aspirations, beliefs, values and ambitions. He (it is usually a man) must become a mythical figure by transcending and redefining the limits of reality as imagined by his fellow beings. By the sheer example of his achievement he has stretched the boundaries of the Universe, taken destiny into his own hands and overcome impossible obstacles and limitations.

This form of transcendence cannot remain immersed in Nature, lose itself in the Universe or unite with the mind of God: it must feed back into society by the power to shape history and inscribe itself in people's minds. In many mythical tales, from Homer to Dante, this was especially the work of the *poet*. For most practical purposes it was he who bestowed the hero with immortality. For instance, on his way through the cycles of Hell, it is the poet Virgil who rescues Dante from his own ignorance and sin, and who guides him on his journey towards God. The storyteller has the power to draw together life's *boundary conditions*: contingencies and accidents, catastrophes and victories, individual and collective destinies, tragedies, suffering, inhuman efforts and ambitions, justice and injustice, belonging and community. The meaninglessness and meaningfulness of it all can be brought within the fold of the Cosmos through a singular, beautifully told story. Although that may remain as true today as in ancient times, the coordinates and forms of cosmic storytelling have greatly changed. *Technoscience* emerges today as the utmost resource for projecting reality. Technoscience pushes hard at the boundaries of reality with promises to create new forms of life from a computer chip, or to reset the global climate system. It also challenges long-standing scientific storytelling rules: not to exceed the empirically verifiable; to retain a skeptical attitude; not to commit to single interests because they will corrupt objectivity (Merton 1973). In many ways, technoscience has wrought free of these limitations. At the same time it strongly relies on them to bestow legitimacy and credibility on its enterprise. What it lacks in scientific rigor it compensates for by re-directing its resources towards what *will* be: new techno-forms of life, new futures. Proponents of prolonged life do not hesitate for long to call upon religious and transcendent dimensions to pursue their business more efficiently. The wish to push away the specter of death is intimately connected with the desire to conquer the future, to get there before and above all others. *Speed* has become of utmost importance, nowhere more clearly seen than in transhumanist models of exponential growth: history is imagined as the pace with which change accelerates through new and game-changing technologies. We will soon arrive at a point in time, so the story goes, where evolution itself will pass through the loophole of technological change, issuing in an explosion of intelligence, the universe itself gaining consciousness and waking up: the Singularity.

It is not easy to write critically yet transparently about phenomena of the present, even less so when they lean towards the extreme. In many respects this chapter remains an essay: a preliminary attempt at mapping some issues and perspectives relevant for understanding engineered immortality in a wider cultural context. To gain some perspective, I start by making a rather lengthy detour to some of the historical origins, or cosmic coordinates, of socio-technical immortality: religion and cosmology, science and nationalism. I shall focus on a few outstanding individuals who succeeded in making themselves immortal. The cases are not intended to be biographical, but are instantiations of emblematic actions situated at defining moments and critical junctures of history. In the final part of the text I return to transhumanism and I ask what perspective could be gained by such a detour.

Variations of socio-technical immortality

Most religions articulate the human condition as one of finite existence within an infinite cosmological order. Many also hold out the promise or possibility of overcoming man's finitude and loneliness by merging with the infinite cosmological order, and so achieving immortality. For instance, the teaching of Hermes Trismegistus has it that "Death does not exist, and man never steps out of universal life; nevertheless, conscious immortality must be gained by each individual for himself" (Hermes in Blavatsky & Judge 1927). According to Hermes, then, men are already living immortal lives. The difficult task is to realize this immortal nature, and so liberate oneself from the eternal absurdities and contingencies to which mortal life is subject. This could only be done through extraordinary effort: "Oh, men, live soberly. Win your immortality" (*ibid.*). Hermes teachings can also be found in Greek philosophy and numerous Christian writings. In Plato's *Republic*, Socrates makes the similar claim that the soul is immortal and cannot be destroyed by its own wickedness. This prepares the ground for the argument that goodness is sought for its own sake, not for its consequences. Still, says Socrates, the good man is also rewarded by society in this life, and the bad man is punished. But the cycle of life does not stop there. Punishment and reward is handed out with even greater force and consequence after death:

> He pays, therefore, tenfold retribution for each crime, and so for instance those who have been responsible for many deaths, by betraying state or army, or have cast others into slavery, or had a hand in any other crime, must pay tenfold in suffering for each offense. And correspondingly those who have done good and been just and god-fearing are rewarded in the same proportion (Plato 2007).

Within the cosmos of the Greek City State, then, good and rightful action could be seen as a goal in itself. But it made little sense to separate efforts to be a good person from the wider mores, aspirations and lives of one's fellow citizens. Thus life in the City State extended indefinitely towards a larger Cosmic whole, within the physical world, between the heavens and the lower worlds. No decisive boundaries existed between these; they were porous. For instance, the epic of Odysseus describes his efforts to return home. It includes his journey into the Underworld, where he meets the dead. He is helped by the advice of the gods, among them Circe, the daughter of Helios the Sun God. The journey takes place among Sirens, monsters, witches, gods and animals, and Odysseus's men are turned into swine and punished for their sins. The world of Odysseus could not be neatly separated into a "good" and a "bad" part; these co-existed, and efforts to be a good person or reach immortality would have to take this state of affairs into account. Efforts to rise above the forces of the Cosmos would lead to ignorance, *hubris*. This multi-layered, vertically ordered Cosmos did not disappear with Christianity, but separations between the spheres gradually tightened. In Christianity there was still the upper world of the heavens; there was the human world, and there was the lower world of suffering, chaos and disorder (Hell). But the lower and upper worlds were gradually relegated from experience, increasingly the exclusive domain of priests.

From the sixteenth century onwards this development was helped by a grand-scale reconfiguration of the Cosmos in the name of the new mathematical sciences, what Weber termed the "disenchantment of the world". Mathematics, geometry and the new experimental sciences provided the groundwork for modern cosmology and served as main resources for a slow but powerful upheaval: the *Universe* came to be seen as dead, mechanistic matter following the causal laws of nature (Shapin 1998). Galileo Galilei not only depicted the book of nature as written in the language of mathematics: he effectively *wrote* the book. Extension, breadth and depth, the three-dimensional space of geometric systems: quantifiable properties were elevated to the hardest and most indubitable forms of reality. Other modes of perception, such as the emotions and sensual impressions through which the world could be poetically, if not scientifically, grasped, were downgraded to illusion. The Universe became an extended surface, re-configured through the lenses of geometric, three-dimensional space. This opened up the perspective for new eternities, new immortalities: Galileo famously turned his gaze towards the heavens as the ultimate horizon. In the process he transcended his times a great many times: "The universe which I with my astonishing observations and clear demonstrations had enlarged a hundred, nay, a thousand fold beyond the limits commonly seen by wise men of all centuries past" (Galilei in Zax 2009). Through geometry and through tuning his telescope towards the heavens, he powerfully re-directed the gaze of those to come after him. But he also unashamedly inscribed the Universe into the *social world* in which he was living. In an unprecedented PR stunt, he used his discoveries to gain patronage from the Medicis, for centuries one of the most powerful families in Tuscany and well connected to the Catholic Church (with which Galileo had a more than difficult relation). Galileo proceeded to elevate the Medici name to eternity by naming a group of stars after one of their sons. This, according to him, was the way to "preserve from oblivion and ruin names deserving of immortality" (*ibid.*).

His telescope and his mathematics came to be seen as the most objective, highest expressions of pure observation that existed. They also provided the lenses through which the universe was re-configured, a spiritual and societal upheaval of cosmic dimensions, and so hardly free from interest. The telescope was a highly appropriate symbol for the emerging concept of *time* as well: a view of the cosmos stretching in a straight line towards a distant final point, *eternity*. Because the material world had been re-imagined through mechanics and mathematics, it came to be seen as without intrinsic value, purpose or goal. At the same time Galileo himself remained a devout Christian. Transcendence and freedom were squeezed into a parallel dimension, existing alongside the physical universe in a strained dualism: supposedly free from the constraints of the physical, yet inscribed with its form. Philosophers like Descartes and Kant were hard at work to align the new sciences with Christianity. Two things fill me with awe, Kant said: *the starry heavens above and the moral law within*. In this way he sought to re-define (and rescue) both religion and science, and to re-align human finitude (morals) with the whole, the endless universe of cause and effect. Religion was possible, Kant said, but only within the limits of reason; the senses were not to be trusted nor relied upon as something other than sources of data acquisition. Immortality was separated from the realm of the real, and turned into something approaching an extra-terrestrial principle: something to be hoped for, but not a true goal of moral action. Within the Greek cosmology time had been circular, and eternity and immortality were never far off. But with the coming of "empty homogenous time" (Walter Benjamin),[29] the sources of transcendence were displaced: now existing outside of horizontal, secular time, they could only be revealed at the end of history or of individual life. It was not so much that the religious disappeared; it was displaced towards the future.

Even though the gods, the heavens, and spirits were expelled, human aspirations and needs for transcendence and belonging were not. Instead they found new expressions, re-located and re-imagined coordinates of self, nature and society. Other forms of social life emerged as the old world gave way to the new. Dynasties and kingdoms tuned into a cosmically ordered universe were replaced by nation states and secular ideologies. In their place appeared a phenomenon that was partially opposed to the universe as described by modern science: nationalism. Says Benedict Anderson.

> With the ebbing of religious belief, the suffering which belief in part composed did not disappear. Disintegration of paradise: nothing makes fatality more arbitrary. Absurdity of salvation: nothing makes another style of continuity more necessary. What then was required was a secular transformation of fatality into continuity, contingency into meaning (Anderson 1983: 11).

Anderson does not ascribe "the ebbing of religious belief" to nationalism; if anything it was a response to an emerging vacuum. To some extent, this vacuum appeared with Galileo and the new sciences. But it was also instigated by the increasingly important role of modern ideologies, on the side of both the laboring and bourgeois classes: "[…]

[29] Through which Benjamin referred to time as mathematical and standardized, and so always and anywhere the same. This, according to him, was how modern reformers saw history and progress.

neither Marxism nor Liberalism are much concerned with death and immortality" (*ibid.*). Neither, we may add, is Darwinism: the process of evolution, although now infused with forms of life, by and large confirmed the Galilean image of nature as a random process without intrinsic purpose. *Fame*, even more than in previous times, provided a way of rising above the ordinary, not the least through improved means of communication and information: national newspapers, novels, maps and museums. Through such media the extraordinary individual could enter the minds of his fellow citizens with much greater symbolic force and speed. Modern-day secular immortality became possible, adapted to the nation state and national consciousness. Living in the minds of his fellow citizens, if not in the flesh, the famous person's memory could be handed down infinitely through the generations.

In Norwegian history two mythical scientist-explorers stand at the threshold of modernity: Fridtjof Nansen and Roald Amundsen, conquerors of the Arctic and the South Pole. At a time of liberation from Sweden these two men stood out as great symbols of newly gained national sovereignty. They eminently embodied the belief in progress and science, and the need for man to conquer nature in the name of the nation. The very name of Nansen's ship expressed just this: "Fram", in English "Forward". The forward-moving momentum was intrinsic to Nansen's explorations: as he set out to reach the North Pole, he described "[...] how well I know this feeling, from each time I set out, and the way back has been cut off [...]" (Nansen 1897, my translation). Although he did not actually make it to the North Pole, the expedition across the Arctic ice sheet was hailed as a huge success. "Fram" ventured further north than any ship before it.[30] Nansen forever inscribed himself in the Polar North, thereby also national consciousness; not only by physically putting it under his feet, but also by the way he documented the deed. Shortly after his return, he wrote a popular account, thanking the Norwegian Parliament, people and King. But the book was also one of meticulous description, using insights from a number of sciences. Between them, Nansen and his men covered disciplines such as botany, geography, meteorology, medicine (Nansen was an early theoretician of the nervous system) and astronomy. The findings and observations made during the expeditions into the Arctic became the basis for numerous scientific publications.[31] In the popular account Nansen's narrative was kept in a positivist and minimalist descriptive language. He did not seem to give in to airy speculations or metaphysical conjecture, and ended his narrative on scientific issues about geography, the movements of the ice, meteorology, the northern lights, animal life, and how to

[30] Later on, "Fram" also carried Amundsen on his expedition to conquer the South Pole.

[31] In addition, Norwegian newspapers eagerly covered, to some extent also financed, his expeditions. Nansen's "follower", Amundsen, the first to set foot on the South Pole (also with "Fram") was hired as a correspondent for Aftenposten.

navigate ships in polar areas. However, upon his homecoming he also recalled the North as follows: "The ice and the Northern full moon with all the longing of the night was only a remote dream, from another world, a dream that lived and faded [...] But then [...] what would life be, without these dreams?" (*ibid.*). This longing and the drive that sustained it were also mobilized for the sake of another struggle, the liberation of Norway from Sweden. As he returned from the Arctic he noted that: "One could feel the pulsation of power and life of this people. In the distance I saw the future rising, great and prosperous, when the powers now suppressed wrench loose and liberate themselves" (*ibid.*). Transcendence was never far from Nansen's efforts: the dreamlike recollection of the distant and other-worldly North there to be conquered; the power of the Norwegian people to grasp destiny and create its own future. These were distinct phenomena. Still, through his actions he brought them together in ways that still linger deep in national consciousness.

Today, of course, much of the historical backdrop and motivation has faded. Obsessive individuals will walk alone to the North Pole as a matter of personal gratification. Such deeds no longer connect with a strong collective feeling or command much respect. It seems that people have already set their feet just about everywhere on the planet. If nationalism is not an option, and if all territories have been discovered: where to direct the drive for transcendence, the race for the future? Clearly, transhumanism is far from the only answer to that question, nor even an especially logical one. But it does, as we shall see, announce itself as one powerful response, and one that is heavily inscribed in the ever-more nervous pace of accelerated progress and growth of post-millennium Western societies. More than anything else, this new space of opportunity was opened up by the discovery of what Claude Bernard called the internal milieu: the internal life of the human body as a new horizon for experimental intervention through science and technology.

Engineered immortality

Transhumanism in its present form only emerged along with the large-scale successes of the experimental life sciences in the first part of the twentieth century. Most of its originators were not original inhabitants of the new experimental systems, as were scientist heroes such as Claude Bernard, Louis Pasteur or Alexander Fleming. Rather, they were a heterogeneous bunch of socio-technical visionaries, mainly from the UK, who mobilized the insights of the laboratories in creative ways. A characteristic trait was their capacity to cross the borders between the arts and the sciences, and to extract from these different worlds powerful visions of *future forms* of technological life as transformed through science and technology. Famous examples include Julian Huxley (scientist and founder of UNESCO), H.G. Wells ("father of science fiction"), and J.D. Bernal (scientist, writer and political activist). These men conceived most of the ideas central to today's transhumanist movement: space travels and mechanical men capable of inhabiting outer space; man-machines and cyborgs, perfect control with emotions, life extension and science-based immortality. Central to their efforts was the articulation of an *ideology of extreme progress* (Coenen 2007), imagined to bring the whole of humanity to a new level. Indeed, human nature itself was seen as the next frontier for exploration:

> We have pretty well finished the geographical exploration of the earth; we have pushed the scientific exploration of nature, both lifeless and living, to a point at which its main outlines have become clear; but the exploration of human nature and its possibilities has scarcely begun [...] The zestful but scientific exploration of possibilities and of the techniques for realizing them will make our hopes rational, and will set our ideals within the framework of reality (Huxley 1957).

Following WWII and the gradual migration of brainpower, technological and political hegemony away from Europe, the same ideas took hold across the Atlantic with California as the great stronghold. Other forerunners of today's movement include the so-called Extropians[32] and thinkers such as Max More and Fereidoun M. Esfandiary (also known as FM-2030). Present-day transhumanism is also intrinsically linked with the broader cultural trend of post-humanism. Post-humanism, which emerged in the early 1980s, can be characterized by its celebrations of hybridity: especially cyborgs and increased man/machine interactions, but also information networks and sensors attached to the human body, prostheses, drugs and medications. Post-humanism was, and remains, highly diversified and without a strong center, message or unifying idea (perhaps apart from the image of the cyborg itself). No clear directionality emerged, and one would be hard pressed to call it a movement. It enjoys strong intellectual ties to post-modernism, with its general rejection of the idea of progress. But it also contained the seeds of what would later emerge as transhumanism, converging around a set of ideas sharing in the ideology of extreme progress. Some such ideas were associated with "space activism", i.e. various initiatives to promote space exploration, a specific goal being colonization and the creation of man/machines capable of inhabiting it. Other ideas included cryonics, the freezing of bodies for the sake of future downloading of consciousness, and up- and down-loading of consciousness to computer networks. Among the driving personalities are nanotech visionary Eric Drexler, the roboticist Hans Moravec and the artificial intelligence pioneer Marvin Minsky (STOA 2009). In terms of official politics, the strongest expression of the

[32] A central claim of extropians being that science and technology have opened up the prospects for people to live indefinitely.

transhumanist movement may have been the so-called NBIC initiative, instigated by Michael Roco and William Bainbridge (Roco & Bainbridge 2002). In the early 2000s, they were at the cusp of efforts to establish converging technologies as a main research program in the US. Both were working for the National Science Foundation (NSF), Roco originally as an engineer and manager, Bainbridge a sociologist of religion. The project mobilized resources from powerful US institutions, such as NASA, the NSF and the Department of Commerce. Other supporters included major actors of the ICT industries such as Intel and Google. After a few years the project was abandoned, seemingly too un-scientific and speculative to be touched by official sanction and support. But such entanglements do say something about the extent to which transhumanism intersects with powerful contemporary trends, compressed by the ideology of extreme progress and a strong belief in the capacity of science and technology to provide solutions to most pressing problems.[33]

Hence, transhumanism is more than anything an ideological phenomenon thriving on the fringes of technoscience. Especially powerful is the idea that sciences and technologies will converge.[34] With their powers immensely enhanced, they can be turned towards solutions for most problems facing humanity, including mortality. In 1993 the mathematician and science fiction writer Vernor Vinge made popular the idea of *the Singularity*. The Singularity refers to a transformative point in time, then predicted by Vinge to occur between 2005 and 2030. Vinge described the Singularity as an "intelligence explosion", resulting from ever-faster implementations of intelligence to an ever-increasing number of processes, and finally to evolution itself:

> We humans have the ability to internalize the world and conduct "what if's" in our heads; we can solve many problems thousands of times faster than natural selection. Now, by creating the means to execute those simulations at much higher speeds, we are entering a regime as radically different from our human past as we humans are from the lower animals (Vinge 1993).

The idea implies a self-accelerating process culminating in large-scale transcendence: evolution *is* development and progress, and has been going on for millions of years. However, if humans think faster than evolution, and if computers can greatly enhance that process, then evolution itself may become intelligent: the universe may wake up, and evolution becomes a conscious process. Artificially created intelligence will

[33] Although abandoned, the NBIC initiative would find its way into other institutions, such as the Immortality Institute and the Singularity University.

[34] For the sake of comprehensiveness: there are varieties of transhumanism not covered in this chapter, some of which significantly deviate from those described here. For instance, humanist and liberal varieties can be found. These, however, have not enjoyed the same institutional influence as the mainstream version.

also have reached the level where it may self-consciously enhance itself, and so the process is one of an ever-faster enhancement-loop. At this point there is of course no guarantee that the artificial intelligence (AI) will be beneficent towards humans, and this was the reason for the sub-title of Vinge's article: *How to Survive in the Post-Human Era*. Humans could become the weaker link in the evolutionary chain, and so be counted out of progress and evolution. Taken to its extreme, the idea of the Singularity may denote the utter meaninglessness of techno-science as a human endeavor: humans making humans superfluous.

The inventor, entrepreneur and futurist Ray Kurzweil has further expounded on the Singularity, and taken it in a seemingly more optimistic direction. Kurzweil (2005) describes history as one of constantly occurring technological revolutions, or paradigm shifts: from the first agricultural settlements through roads through the printed word through air travel to computers. Paraphrasing Moore's law, which (roughly) states that computer processing power doubles every two years, Kurzweil sees history as developing through ever-faster growth, or "accelerating returns". Progress, he states, is not linear; it is exponential. By approximately 2045, evolution will pass through that magic point in time where progress within a number of technology areas, especially nanotech, robotics, artificial intelligence and computing, will converge.[35] The Singularity will radically transform humanity, the Earth, everything we know about it and the ways we inhabit it. Beyond this loophole of total transformation we know nothing, yet Kurzweil maintains the general consequences to be beneficial and worth striving for:

> These technological revolutions will allow us to transcend our frail bodies with all their limitations. Illness, as we know it, will be eradicated. Through the use of nanotechnology, we will be able to manufacture almost any physical product upon demand, world hunger and poverty will be solved, and pollution will vanish. Human existence will undergo a quantum leap in evolution. We will be able to live as long as we choose. The coming into being of such a world is, in essence, the Singularity (Kurzweil 2006).

Among the main benefits of the Singularity, Kurzweil predicts, is the possibility for directly re-programming the human organism so as to delay or avoid death altogether. Here, transcendence is imagined as, first, proceeding through gradual "re-programmations" of the human body, be that through interventions with the body itself or by merging with machines and artificial intelligence. As an example of the first, life extension should be possible through genetic engineering or through biochemical and dietary measures. Kurzweil himself takes approximately 200

[35] Moore himself has decreed Kurzweil's writings as unscientific speculations, as have a number of other scientists.

supplement pills a day, follows a strict diet, drinks ionized water and meticulously monitors and records his health data from day to day. There is also biomedical and biochemical research going on, such as that of Aubrey de Grey, into "delayed senescence" (delayed aging). But Kurzweil ultimately believes that human DNA and biological evolution has handed humanity poor cards, rendering the body a "second-class robot". Decay may be unavoidable. This could partially be fixed by nanotechnology, -enabling us to re-engineer the body molecule by molecule. Even more promising, though, is the coming robotics revolution, by which Kurzweil primarily means an explosion in artificial intelligence. Because human beings are essentially information (DNA), the patterns, knowledges and skills that make up each single individual can be "backed up" and stored in computers. Indeed, Kurzweil himself has already reserved a space for his body to be cryo-preserved, in case he is not able to sustain his own physical existence until the coming of the Singularity (Kurzweil was born in 1948, so with the coming of the Singularity, as predicted by himself, he will be 97 years old). The Singularity, thus, entails the total overcoming of most limitations still believed to be fundamental:

> This merger of man and machine, coupled with the sudden explosion in machine intelligence and rapid innovation in gene research and nanotechnology, will result in a world where there is no distinction between the biological and the mechanical, or between physical and virtual reality (Kurzweil 2006).

Once overcome, human beings may take on God-like capacities. Kurzweil's final remark in the movie made about him (*Transcendent Man*, Ptolemy 2009) goes: *"So you might ask 'Does God exist?' Well, I would say not yet"*. The extent to which Kurzweil conflates god-like powers with computing capacity is also tellingly illustrated by another scene in *Transcendent Man,* where he is standing in front of the ocean, seemingly contemplating it. In awe and wonder he exclaims: *imagine the computing capacity that goes into all this movement!* Bearing in mind the previous descriptions of transcendence, what stands out is how the future appears utterly locked into a techno-scientific materialist projection. That is, in Kurzweil's universe there is no reality outside of information and the material, represented by nano-molecules and genes that can be manipulated at will. This process seems utterly incapable of conceiving anything outside of itself, and so also incapable of dreaming up anything but its own in(de)finite extension. There seems to be no way out of the process or the logic. *Transcendent Man anno 21st century* has gotten himself into a desperate position: the end of life as the ultimate disappointment. Death might appear as a failure to compute, as syntax error.

Concluding: changing coordinates of transcendence

In this chapter I have leaped through a number of epochs and some of their predominant ways of imagining immortality. I have made huge generalizations, and many forms of immortality have not even been mentioned. I have focused on what I have called "socio-technical immortality". Clearly, a focus on immortality as seen by mystics, by everyday persons or by more humble-minded scientists than those described here would have resulted in very different stories to be told. In this text, the "immortal individuals" come across as transitional figures, as outstanding individuals situating themselves at the threshold of new eras and imprinting their marks on the new ages. However, the main point was to gain some perspective on a phenomenon of the present, transhumanism. Let it be said right away: I do not believe Ray Kurzweil to be of the same caliber as a Galileo, Socrates or Nansen. But I do believe him to be a transitional figure. And the *scale* of his claims puts him in this league. One may deride those claims as unscientific ramblings of a lost soul. But the kinds of arguments he uses, and the kind of dreams that he dreams, are more widely spread among those on the cusp of innovation and development than frequently recognized. This speaks about a fundamental problem on the intersections of technoscience, power and politics: Kurzweil himself has insulated himself from criticism in the present and seems to be living and speaking from a future point, from above and beyond his contemporaries. Counter-arguments can be discarded with as relics of the unenlightened past. In this sense, at least, he has transcended his times. There are numerous reasons (historical, scientific, religious, etc.) to believe that most transhumanist claims are wrong, indeed bound for tragedy. Its most ardent promoters would do wise to consider the Greek concept of *hubris*, as when Icarus flew to close to the sun and his wings (of wax) melted. They should also heed the story of the fortune-tellers from Dante's Inferno: too eager to foresee the future, they were punished by having their heads put on backwards, forever condemned to look at the past even as they moved forwards.

The most striking contrast between transhumanism and earlier imaginations of immortality is how radically "eternity", or "the universe", has been transformed. It has ceased to be something "out there", as seen by Galileo through his telescope or by Nansen dreaming of the North. Instead, it is transformed into extreme immanence in the form of computing capacity, the ability to process unfathomable amounts of information simultaneously and at an incredible speed. This capacity of computers is also a radical constructivism, imagined as extreme computation feeding directly into the material universe. The utmost horizon of reality is encountered at the molecular and nano levels, as eternally malleable matter to be manipulated at will. As man becomes all-powerful (or its opposite), there are no longer any boundaries to what appears as possible, feasible, permissible or morally justified. The *social* horizon has by and large disappeared: transhumanists will occasionally speak about national competitiveness (mainly the US) or Humanity, but the main frame of reference remains the individual, and his and her choice (hence the market remains an important point of reference). The main driving force of this isolated individual, it seems, is pure survival for its own sake. Apart from an indefinite quantitative extension of life, there is almost a complete lack of vision of the good life, fundamental values or community. Perhaps if some fundamental values and community existed as *real presences* in life, death and disease could have a meaning and become bearable (as it was for most people in most periods of history). Lacking this, one (extreme) goal of the individual becomes that of driving out all discomfort, sickness and disease, to push these as far into the distant future as possible, where hopefully they will disappear altogether. For non-believers, it is denial of life's basic limits, hence a form of extremism.

Alternatively, the versions of transhumanism here dealt with are not expressions of real belief, but rather cynical, desperate or extremely naïve measures to regain control and a sense of purpose: re-conquering a future that seems to diminish with each passing day. It is indeed remarkable that, at a time when the general belief in Western progress and growth is at its lowest, the ideal of extreme progress fights its way back. It was not long ago that one famous cultural analyst proclaimed the "end of history" and the victory of the Western model. But right after that Western capitalism was hit at its core: first the Twin Towers collapsed in flames, then the economic crisis hit (the consequences of which we can still barely begin to grasp). This decline is no short-term trend: it has been going on since the early 1970s, when transhumanism in its present-day form began to grow. This was the time when the revolutions in genetic engineering and computer science gained traction, whereas traditional industries went into decline. Since then, pressures have been mounting for science and technology to inject the economy with renewed power and initiative. Genomics and computers are real game-changers. But they do not easily translate into growth, clean energies or improved health, as long-since proclaimed and expected. The metaphor of computer-driven, accelerated growth, then, is mainly referring to itself, and not to something outside it. This was clearly illustrated by the scene from *Transcendent Man*, where Kurzweil is standing in front of the eternal ocean, seemingly unable to appreciate anything but its digital quantity.

I have told a story of a certain co-production of the cosmos, science and society. Indeed, at times the story comes too close to the story of extreme progress that I wanted to criticize: Galileo laid down the fundamental coordinates, later generations filled in the spaces as best as they could. In certain ways, it now seems that we have exhausted the potential of this cosmology, and the "long march of the West" (Taylor 2004) is at a turning point. The knee-jerk reaction to the same conclusion is to push even harder for business as usual, for *much* more of the same. But perhaps this is just a symptom of a certain exhaustion, and perhaps one does not have to change one's perspective that much for new meanings and horizons to appear. And perhaps we may even stay with the computer and with information, but inscribe them in a universe of *tangible* change and aspiration. 2011 was the year when millions of people across the globe used computers and social media to mobilize protests against the injustices of globalization, in many ways the true face of the progress celebrated by Kurzweil and his peers. In spite of obvious dangers and pitfalls: there lie the seeds of another form of transcendence, one that has humanity as its ultimate horizon.

Bibliography

Anderson, B. 2006 [1983]. *Imagined Communities.* London/New York: Verso.

Blavatsky, H.P. & W. Judge. 1927. Ancient Landmarks XX: Egyptian "Immortality". THEOSOPHY 15 (11), September. Retrieved from http://iml.jou.ufl.edu/projects/fall04/hilton/underworld.htm.

Coenen, C. 2007. Utopian Aspects of the Debate on Converging Technologies. In G. Banse, I. Hronszky & G. Nelson (eds.). *Assessing Societal Implications of Converging Technological Development.* Berlin: Sigma. 141-172.

Huxley, J. 1957. Transhumanism. Retrieved from http://www.transhumanism.org/index.php/WTA/more/huxley.

Immortality Institute. 2004. Introduction. In Immortality Institute (ed.). *The Scientific Conquest of Death. Essays on Infinite Lifespans.* Retrieved from http://www.imminst.org/book.

Kurzweil, R. 2005. *The Singularity is Near.* London: Penguin Books.

Kurzweil, R. 2006. Reinventing Humanity: The Future of Machine-Human Intelligence. *The Futurist.* 40(2): 39-46.

Latour, B. 1993. *We Have Never Been Modern.* New York: Harvester Wheatsheaf.

Merton, R.K. 1973. The Normative Structure of Science. In R.K. Merton (ed.). *The Sociology of Science: Theoretical and Empirical Investigations.* Chicago: University of Chicago Press.

Nansen, F. 1897. *Fram over Polhavet. Den norske polarfærd 1893-1896.* Kristiania: Aschehoug.

Plato 2007. *The Republic.* 2nd edition. London: Penguin Classics.

Ptolemy, B. 2009. *Transcendent Man.* Documentary. Docurama.

Roco, M. & W. Bainbridge (eds.). 2002. Converging Technologies for Improving Human Performance. Retrieved from http://wtec.org/CT/Report/NBIC_report.pdf.

Shapin, S. 1998. *The Scientific Revolution.* Chicago: Chicago University Press.

STOA 2009. EUROPEAN PARLAMENT. *Science and Technology Options Assessment.* HUMAN ENHANCEMENT STUDY. Brussels. Retrieved from http://www.europarl.europa.eu/stoa/default_en.htm.

Taylor, C. 2004. *Modern Social Imaginaries*. Durham: Duke University Press.
Vinge, V. 1993. The Coming Technological Singularity: How to Survive in the Post-Human Era. Retrieved from http://www-rohan.sdsu.edu/faculty/vinge/misc/singularity.html.
Zax, D. 2009. *Galileo's Vision*. Smithsonian magazine, August.

Chapter 10: What should be the role of religion in science education and bioethics?

Michael J. Reiss

> [...] discussion of religious beliefs between a believer and a non-believer can seem superficial to the former and frustrating to the latter.
>
> *Hinde 1999*

In much of Europe the importance of religion has been waning for at least a hundred and fifty years, probably much longer. However, somewhat to the surprise of many people, religion continues to be important in much of European social life and politics. In certain areas, including aspects of education and bioethics, it may, if anything, be increasing in influence in some countries. This chapter will concentrate on the relevance of this to science education and bioethics. What role, if any, should religion play in these matters and how should we decide this question? I shall argue that science education and bioethics need to respond quite differently to the issue of religion, and that the reason for this difference stems from the different aims of science education and bioethics and the different ways in which the two disciplines arrive at their claims about valid knowledge.

The role of religion in science education

For many science educators, whether or not they have any religious beliefs themselves, the relationships between science and religion, i.e. the "science/religion issue", appears somewhat outside the scope of science education. However, a range of factors, including a greater awareness of the benefits of dealing explicitly in the school classroom with the nature of science and the increasing influence of creationism in schools, suggests that this perspective may be too narrow (Reiss 2008).

The function of school science education is principally to introduce learners to the methods that the sciences use and to the different forms of knowledge that the sciences have produced. While historians tell us that what scientists study changes over time, there are reasonable consistencies:

1. Science is concerned with the natural world and with certain elements of the manufactured world – so that, for example, the laws of gravity apply as much to aeroplanes as they do to apples and planets.

2. Science is concerned with how things are rather than with how they should be. So there is a science of nuclear fission and *in vitro* fertilization, without science telling us whether nuclear power and test-tube babies are good or bad.

The argument in favor of including religion in science education is then a very specific one: aspects of religion should be included if they help students to learn science better. (Precisely the same argument holds, I would argue, for teaching science students about history: this too should be done if it helps students to learn science better.) So, under what circumstances might the learning of science be helped by a consideration of religious issues? Perhaps the most obvious instance is when teaching the topic of evolution to students who are creationists or, at any rate, have creationist sympathies.

The importance of creationism for science education

Creationism exists in a number of different versions, but something like 50% of adults in Turkey, 40% in the USA and 15% in Norway reject the theory of evolution: they believe that the Earth came into existence as described by a literal (fundamentalist) reading of the early parts of the Bible or the Qu'ran and that the most that evolution has done is to change species into closely related species (Miller *et al.* 2006). For a creationist it is possible, for example, that the various species of mice had a common ancestor but this is not the case for mice, squirrels and horses – still less for monkeys and humans, for birds and reptiles or for fish and pine trees.

Until recently little attention has been paid in the science classroom to creationism. However, creationism appears to be on the increase, and there are indications that there are more countries in which schools are becoming battlegrounds for the issue. For example, while the USA has had several decades of legal battles about the place of creationism and (more recently) intelligent design in schools (Moore 2007), school-based conflicts over these issues are becoming more frequent in a range of other countries (e.g. Graebsch & Schiermeier 2006).

As a result, there has been a growth in the science education literature examining creationism (e.g. Jones & Reiss 2007, Reiss 2011). Most of the literature on creationism (and/or intelligent design) and evolutionary theory puts them in stark opposition. Evolution is consistently presented in creationist books and articles as illogical (e.g. natural selection cannot, on account of the second law of thermodynamics, create order out of disorder; mutations are always deleterious and so cannot lead to improvements), contradicted by scientific evidence (e.g. the fossil record shows human footprints alongside animals supposed by evolutionists to be long extinct;

the fossil record does not provide evidence for transitional forms) and the product of non-scientific reasoning (e.g. the early history of life would require life to arise from inorganic matter – a form of spontaneous generation rejected by science in the nineteenth century. Radioactive dating is said to make assumptions about the constancy of natural processes over eons of time whereas we increasingly know of natural processes that affect the rate of radioactive decay), and evolution in general is portrayed as the product of those who ridicule the word of God, and a cause of a whole range of social evils – from eugenics, Marxism, Nazism and racism to juvenile delinquency, illicit drug use and prostitution (e.g. Baker 2003, Parker 2006, Watson 1975 and countless articles in the publications of such organizations as Answers in Genesis, the Biblical Creation Society, the Creation Science Movement and the Institute for Creation Research).

By and large, creationism has received similarly short shrift from those who accept the theory of evolution. In an early study the philosopher of science Philip Kitcher argued that "... in attacking the methods of evolutionary biology, Creationists are actually criticizing methods that are used throughout science" (Kitcher 1983: 4-5). Kitcher concluded that the flat-earth theory, the chemistry of the four elements and mediaeval astrology "... have just as much claim to rival current scientific views as Creationism does to challenge evolutionary biology" (*ibid.* 5). An even more trenchant attack on creationism is provided by geologist Ian Plimmer whose book title *Telling lies for God: Reason vs creationism* (Plimmer 1994) indicates the line he takes.

The scientific worldview is materialistic in the sense that it is neither idealistic nor admits of non-physical explanations (here, "physical" includes, as well as matter, such "things" as energy and the curvature of space). There is much that remains unknown about evolution. How did the earliest self-replicating molecules arise? What caused membranes to exist? How key were the earliest physical conditions – temperature, the occurrence of water and so forth? But the scientific presumption is either that these questions will be answered by science or that they will remain unknown. Although some scientists might (sometimes grudgingly) admit that science cannot disprove supernatural explanations, scientists do not employ such explanations in their work (the tiny handful of seeming exceptions only attest to the strength of the general rule).

Whereas there is only one mainstream scientific understanding of today's biodiversity, there are a considerable number of religious ones. Many religious believers are perfectly comfortable with the scientific understanding, either on its own or accompanied by a belief that evolution in some sense takes place within God's holding (compass or care), whether or not God is presumed to have intervened or acted providentially at certain key points (e.g. the origin of life or the evolution of humans). But many other religious believers adopt a more creationist perspective or that of intelligent design (Reiss 2008).

The response of science education to creationism

Given all this, how might raising the issue of religion in science lessons help? Might it not just make the situation even worse? The response by science education to the range of positions held about evolution needs, I believe, to take account of the following (Reiss, in press):

1. Among scientists, the theory of evolution is held to be a robust, well established and, at its core, a scientifically uncontroversial theory.
2. Within biology, evolution occupies a central place. There is much in biology that has been discovered and can be studied without accepting the theory of evolution, but an evolutionary framework is what enables biologists to provide coherence to the diversity of life that we see around us and to situate today's life in an historical context.
3. In common with many scientific theories, evolution is not easy to understand. It has contra-intuitive elements and, in addition, is actively rejected by many people for religious reasons.

Few countries have produced explicit guidance as to how schools might deal with the issues of creationism or intelligent design in the science classroom. One country that has is England (*ibid.*). In the summer of 2007, after months of behind-the-scenes meetings and discussions, the then DCSF (Department of Children, Schools and Families) Guidance on Creationism and Intelligent Design received Ministerial approval and was published (DCSF 2007). The Guidance points out that the use of the word "theory" in science (as in "the theory of evolution") can mislead those not familiar with science as a subject discipline because it is different from the everyday meaning, when it is used to mean little more than an idea. In science the word indicates that there is a substantial amount of supporting evidence, underpinned by principles and explanations accepted by the international scientific community.

The DCSF Guidance goes on to say: "Creationism and intelligent design are sometimes claimed to be scientific theories. This is not the case as they have no underpinning scientific principles, or explanations, and are not accepted by the science community as a whole" (DCSF 2007) and then states:

> Creationism and intelligent design are not part of the science National Curriculum programmes of study and should not be taught as science. However, there is a real difference between teaching "x" and teaching *about*

> "x". Any questions about creationism and intelligent design which arise in science lessons, for example as a result of media coverage, could provide the opportunity to explain or explore why they are not considered to be scientific theories and, in the right context, why evolution is considered to be a scientific theory (*ibid.*).

This seems to me a key point and one that is true for all countries, whether a country permits the teaching of religion (as in the UK) or does not (as in France, Turkey and the USA). Many scientists, and some science educators, fear that consideration of creationism or intelligent design in a science classroom legitimizes them. For example, the excellent book *Science, evolution, and creationism* published by the US National Academy of Sciences and Institute of Medicine asserts "The ideas offered by intelligent design creationists are not the products of scientific reasoning. Discussing these ideas in science classes would not be appropriate given their lack of scientific support" (National Academy of Sciences and Institute of Medicine 2008: 52).

As I have argued (Reiss 2008), I agree with the first sentence of this quotation but disagree with the second. Just because something lacks scientific support doesn't seem to me a sufficient reason to omit it from a science lesson. Indeed, good science teaching typically requires consideration of students' ideas when these do not agree with scientific knowledge. Nancy Brickhouse and Will Letts (1998) have argued that one of the central problems in science education is that science is often taught "dogmatically". With particular reference to creationism they write:

> Should student beliefs about creationism be addressed in the science curriculum? Is the dictum stated in the California's *Science Frameworks* (California Department of Education, 1990) that any student who brings up the matter of creationism is to be referred to a family member or member of the clergy a reasonable policy? We think not. Although we do not believe that what people call "creationist science" is good science (nor do scientists), to place a gag order on teachers about the subject entirely seems counterproductive. Particularly in parts of the country where there are significant numbers of conservative religious people, ignoring students' views about creationism because they do not qualify as good science is insensitive at best (Brickhouse & Letts 1998: 227).

It seems to me that school science lessons should present students with the scientific consensus about evolution and that parents should not have the right to withdraw their children from such lessons. Part of the purpose of school science lessons is to introduce students to the main conclusions of science – and the theory of evolution is one of science's main conclusions. At the same time, science teachers should be respectful of any students who do not accept the theory of evolution for religious (or any other) reasons. Indeed, nothing pedagogically is to be gained by denigrating or ridiculing students who do not accept the theory of evolution.

My own experience of teaching the theory of evolution for some thirty years to school students, undergraduate biologists, trainee science teachers, members of the general public and others is that people who do not accept the theory of evolution for religious reasons are most unlikely to change their views as a result of one or two lessons on the topic, and others have concluded similarly (e.g. Long 2011). However, that is no reason not to teach the theory of evolution to such people. One can gain a better understanding of something without necessarily accepting it. Furthermore, recent work suggests that careful and respectful teaching about evolution can indeed make students considerably more likely to accept at least some aspects of the theory of evolution (Winslow *et al.* 2011).

The role of religion in bioethics

Ethics is the branch of philosophy concerned with how we should decide what is morally wrong and what is morally right. Bioethics is specifically concerned with the application of ethics to the whole of nature, not just to humans (as in medical ethics and much traditional ethics). Much of my argument about the importance of religion holds for all aspects of ethics but I ground what I claim with reference to the particular field of bioethics.

Ethics is a branch of knowledge just like other intellectual disciplines, such as science, mathematics and history. Ethical thinking is not wholly distinct from thinking in other disciplines but it cannot simply be reduced to them. In particular, ethical conclusions cannot be unambiguously proved in the way that mathematical theorems can. However, this does not mean that all ethical conclusions are equally valid. After all, most philosophers of science would hold that scientific conclusions cannot be unambiguously proved, instead remaining as provisional truths, but this does not mean that my thoughts about black holes are as valid as Stephen Hawking's. Some conclusions – whether in ethics, science or any other discipline – are more likely to be valid than others.

One can be most confident about the validity and worth of an ethical conclusion if three criteria are met (Reiss 1999). First, if the arguments that lead to the particular conclusion are convincingly supported by reason. Secondly, if the arguments are conducted within a well established ethical framework. Thirdly, if a considerable degree of consensus exists about the validity of the conclusions, arising from a process of genuine debate.

It might be supposed that reason alone is sufficient for one to be confident about an ethical conclusion. However, there are problems in relying on reason alone when thinking ethically. In particular, there still does not exist a single universally accepted framework within which ethical questions can be decided by reason (O'Neill 1996, Parfit 2011). Indeed, it is unlikely that such a single universally accepted framework will exist in the foreseeable future, if ever. This is not to say that reason is unnecessary but to acknowledge that reason alone is insufficient. For instance, reason cannot decide between an ethical system which looks only at the consequences of actions and one which considers whether certain actions are right or wrong in themselves, whatever their consequences. Furthermore, feminists and others have cautioned against too great an emphasis upon reason. Much of ethics still boils down to views about right and wrong, informed more by what seems "reasonable" than what follows from formal reasoning.

The insufficiency of reason is a strong argument for conducting debates within well-established ethical frameworks, when this is possible. Traditionally, the ethical frameworks most widely accepted in most cultures arose within systems of religious belief. Consider, for example, the questions "Is it wrong to lie? If so, why?". There was a time when the majority of people in many countries would have accepted the answer "Yes, because scripture forbids it". Nowadays, though, not everyone accepts scripture(s) as a source of authority. Another problem, of particular relevance when considering the ethics of contemporary science and technology, is that while the various scriptures of the world's religions have a great deal to say about such issues as theft, avarice, killing people and sexual behavior, they say rather less that can directly be applied to the debates that surround many of today's bioethical issues, for example those involving modern biotechnology (genetic engineering, cloning, stem cells, etc.). A further issue is that we are more conscious nowadays that we live in multicultural or pluralist societies. Within most countries there is no longer a single shared set of moral values.

Nevertheless, there is still great value in taking seriously the various traditions – religious and otherwise – that have given rise to ethical conclusions. People do not live their lives in ethical isolation: they grow up within particular moral traditions and their subsequent ethical views are shaped by those whom they meet, read about or hear. Even if we end up departing somewhat from the values we received from our families and those around us as we grew up, none of us derives our moral beliefs from first principles, *ex nihilo*, as it were. In the particular case of moral questions concerning contemporary biology, a tradition of ethical reasoning is already beginning to accumulate. Many countries have official committees or other bodies looking into the ethical issues that surround at least some aspects of biotechnology. The tradition of ethical reasoning in this field is nothing like as long established as, for instance, the traditions surrounding such questions as war, abortion, euthanasia and trade protectionism. Nevertheless, there is the beginning of such a tradition and similar questions are being debated in many countries across the globe.

What then is the specific place for religion?

In a recent book titled *Dishonest to God: On keeping God out of politics*, Mary Warnock (2010), despite having a certain affection and sympathy for the Church of England, lists many examples where religious arguments have in her view inappropriately been used in parliamentary debates in attempts, some successful, some unsuccessful, to influence national legislation. She concludes: "The danger of religion, any religion, lies in its claim to absolute immutable moral knowledge which, if justified, would indeed give its adherents a special place in instructing others how to behave, perhaps even a right to do so" (Warnock 2010: 165).

My position is close to that of Warnock's but let me finesse it slightly. First of all, we do not need to determine whether or not any particular religion has access to "absolute immutable moral knowledge". As it happens, my understanding of both theology and the human condition is that no one has access to such knowledge. Our concern here is not so much with knowledge as with how one makes practical decisions in a world with a multiplicity of values, religious and otherwise. And here religion has a place at the table. In just the same way as consequentialists have to learn to accept that many deontologists are not going to accept the consequentialist understanding of ethics as being decisive, and vice versa, so ethicists of no religious persuasion need to accept that significant numbers of people have religious beliefs and hold that these beliefs help shape what is deemed morally right and morally wrong.

In this sense, those of no religious persuasion need, I would argue, to take the same sort of account of religious believers as those who eat meat need to take account of vegetarians. We would deem it unacceptable, nowadays, for the authorities in charge of a prison, a hospital or any other residential establishment to fail to provide vegetarian food on the grounds that vegetarianism is unnecessary, a minority lifestyle choice or a fad. In the same way, a secular society that respects its citizens needs to take account of religious views. Of course, precisely the converse holds too. A theocracy that respects its citizens needs to take account of the views of those who have no religious faith or belong to a minority faith.

This may sound rather neat and tidy. How would it work out in practice? Well, in fact it pretty much is working out in practice in a number of countries. In modern democracies we are used to the idea that the best approach to determining what to do when there are deep, genuine differences of opinion (whether in ethics or anything else) is to strive to obtain consensus (Moreno 1995). It is true that consensus does not solve everything. For a start, what does one do when consensus cannot be arrived at? Nor can one be certain that consensus always arrives at the right answer – a consensus once existed that women should not have the vote and that beating was good for children.

Nonetheless, there are good reasons both in principle and in practice for searching for consensus. Such a consensus should be based on reason and genuine debate and take into account long established practices of ethical reasoning. At the same time, it should be open to criticism, refutation and the possibility of change. Finally, consensus should not be equated with majority voting. Consideration needs to be given to the interests of minorities, particularly if they are especially affected by the outcomes, and to those – such as young children, the mentally infirm and non-humans – unable to participate directly in the decision-making process. At the same time it needs to be born in mind that while a consensus may eventually emerge, there is often an interim period when what is more important is simply to engage in valid debate in which the participants respect one another, so far as is possible, and seek for truth through dialogue (cf. Habermas 1983, Martin 1999).

In the case of bioethics, many countries now have well respected bodies that seek to arrive at consensus with regards to contentious ethical issues. Examples include the Norwegian Biotechnology Advisory Board (www.bion.no/) and the Nuffield Council on Bioethics (www.nuffieldbioethics.org/). It seems to me perfectly appropriate that the degree of religious involvement in such bodies should vary from country to country (depending on the extent and depth of religious belief in the population) and from topic to topic (depending on the strength of the connections between religion and the topic in question).

I am well aware that to many with a religious faith this may seem like "selling out". To this objection I would respond as follows. First, it's as good as you are going to get nowadays in an increasing number of countries! Secondly, if a religious viewpoint has sufficient validity, it should be capable of holding its own in arguments with those who have no religious faith. For example, while Roman Catholic arguments about the unacceptability of contraception are very difficult to defend to non Roman Catholics, more broad-based arguments about the sanctity of human life and therefore the unacceptability of euthanasia can receive a more sympathetic hearing among a secular audience so long as "the sanctity of human life" is not seen as a trump card but is translated into religiously neutral language about respect and the protection of the vulnerable. Thirdly, my own reading of the Christian scriptures is that God's nature is such that there are rarely voices from heaven. Usually, determination of what is morally right and morally wrong, while influenced by the reading of scripture and an understanding of the traditions of the church, needs supplanting by broader reflection and study and should be informed, in the case of bioethics, by on-going advances in the biosciences.

A few closing points. One problem with religious viewpoints in ethics is that there are rather a lot of them! A practical consequence of the argument I have advanced is that it may not suffice to have a single religious expert on a bioethics committee. I have occupied such a role more than once and while one can strive to represent the views of absent others, it is better not to have just the one voice. Of course, a bioethics committee, as is the case for almost any committee, cannot grow too large, so it may be necessary to have a system for ensuring that the views of others can feed in in other meaningful ways. But this is simply good practice for dealing with a plurality of viewpoints even if we weren't considering the role of religion in bioethics.

And then there is the objection that the line I have been advancing is a relativistic one that depends on the specifics of history and geography. This is a common objection – not just in theology and bioethics but in other disciplines including science and aesthetics – and a standard response, and one that I hold, is to assert that to deny absolute immutable knowledge is not necessarily to slide inexorably into relativism. One can occupy a middle ground. Indeed, as Parfit (2011) concludes, there are considerable commonalities between the main secular ethical frameworks (Kantian deontology, consequentialism and contractualism) once one gets down to specifics.

Finally, there will be some, who may or may not be atheists, who are not convinced that religion has any role to play in bioethics. Religion, it might be maintained, rests on irrational beliefs in the supernatural and while notions of respect may require us to tolerate such views, nothing should be done that might allow them to influence public policy. It's fine for people to have freedom of expression (e.g. freedom to attend worship) but that is entirely separate from granting religion a public role. If religion were to enjoy such privileges, we would have to extend them to other odd belief systems, such as those who believe they have been abducted by aliens (Clancy 2005) or those who hold that Elvis Presley is still alive (e.g. Brewer-Giorgio 1988, Elvis Is Alive 2012).

There are several reasons why this line of argument does not work. First, the proportion of the population, even in more secular countries, who have some religious beliefs is considerably higher than the proportion of the population who believe in alien abductions or Elvis's longevity. Secondly, religious faith has been around for all of human time whereas conspiracy theories and fads come and go. Thirdly, religious beliefs are often core to a person's being in a way that alien abduction (however upsetting) and Presley mania are but rarely. Fourthly, there is a close connect between many bioethical issues and religious faith which there isn't between bioethical issues and alien abduction or Elvis Presley. Of course, if the state were to set up a publicly-funded museum about aliens, then there might well be a case for granting a voice to those who believe they have experienced such abductions.

Conclusions

The role of religion is therefore, I would argue, different in science education and in bioethics. In science education, a teacher needs to be sensitive to religious objections to aspects of the science curriculum for two reasons: first out of respect for students; secondly, because not to be sensitive is to make learning in science less likely for some students. However, it is not the case that a science teacher should alter the science that is taught because of the religious views of students or anyone else. Scientific knowledge is independent of religious views. In the case of evolution, science teachers may decide not to try to persuade creationist students that they are mistaken but all students, including creationists ones, should be introduced to what science teaches about evolution. At the same time, well-designed examination material should be able to test student knowledge of science and its methods without expecting students to have to convert, or pretend that they have converted, to a materialistic set of beliefs. So, for example, it is appropriate to ask students to explain how the standard neo-Darwinian theory of evolution attempts to account for today's biodiversity but it is not appropriate to ask students to explain how the geological sciences conclusively prove that the Earth is billions of years old.

In bioethics, though, religious views, while they should not have the power that some religious believers would like, nevertheless can, indeed often should, have a place in decision making. A well argued religious viewpoint is neither privileged nor disqualified simply by virtue of its being religious. The same point holds equally for agnostic and atheistic views. In a multicultural society we need to hear a diversity of well argued viewpoints. Much then depends on how those with a religious viewpoint and those without one treat one another and deal with the questions that are being debated, whether in bioethics or elsewhere. When this is done well, all parties can learn from one another without necessarily shifting from their own positions – though, of course, all of us should be open to the possibility of such intellectual growth.

Bibliography

Baker, S. 2003. *Bone of contention: Is evolution true?* 3rd edn. Rugby: Biblical Creation Society.

Brewer-Giorgio, G. 1988. *Is Elvis alive?* New York: Tudor Communications.

Brickhouse, N.W. & W.J. Letts. 1998. The problem of dogmatism in science education. In J.T. Sears & J.C. Carper (eds.). *Curriculum, religion, and public education: Conversations for an enlarging public square*. New York: Teachers College, Columbia University. 221-230.

Clancy, S.A. 2005. *Abducted: How people come to believe they were kidnapped by aliens*. Cambridge, Massachusetts: Harvard University Press.

DCSF 2007. Guidance on creationism and intelligent design. Retrieved from http://webarchive.nationalarchives.gov.uk/20071204131026/http://www.teachernet.gov.uk/docbank/index.cfm?id=11890.

Elvis is Alive 2012. Proof. Retrieved from www.elvis-is-alive.com/.

Graebsch, A. & Q. Schiermeier. 2006. Anti-evolutionists raise their profile in Europe. In *Nature* 444: 406-407.

Habermas, J. 1983. *Moralbewusstsein und Kommunikatives Handeln*. Frankfurt a.M.: Suhrkamp Verlag.

Hinde, R.A. 1999. *Why Gods persist: A scientific approach to religion*. London: Routledge.

Jones, L. & M.J. Reiss (eds.). 2007. *Teaching about scientific origins: Taking account of creationism*. New York: Peter Lang.

Kitcher, P. 1983. *Abusing science: The case against creationism*. Milton Keynes: Open University Press.

Long, D.E. 2011. *Evolution and religion in American education: An ethnography*. Dordrecht: Springer.

Martin, P.A. 1999. Bioethics and the whole: pluralism, consensus, and the transmutation of bioethical methods into gold. In *Journal of Law, Medicine & Ethics* 27: 316-327.

Miller, J.D., E.C. Scott & S. Okamoto. 2006. Public acceptance of evolution. In *Science* 313: 765-766.

Moore, R. 2007. The history of the creationism/evolution controversy and likely future developments. In L. Jones & M.J. Reiss (eds.). *Teaching about scientific origins: Taking account of creationism*. New York: Peter Lang. 11-29.

Moreno, J.D. 1995. *Deciding together: Bioethics and moral consensus*. Oxford: Oxford University Press.

National Academy of Sciences and Institute of Medicine 2008. *Science, evolution, and creationism*. Washington, DC: National Academies Press.

O'Neill, O. 1996. *Towards justice and virtue: A constructive account of practical reasoning*. Cambridge: Cambridge University Press.

Parfit, D. 2011. *On what matters: Volume 1.* Oxford: Oxford University Press.

Parker, G.E. 2006. *Creation: Facts of life – how real science reveals the hand of God.* Los Angeles, CA: New Leaf Press.

Plimmer, I. 1994. *Telling lies for God: Reason vs creationism.* Milsons Point, NSW: Random House.

Reiss, M.J. 1999. Bioethics. In *Journal of Commercial Biotechnology* 5: 287-293.

Reiss, M.J. 2008. Should science educators deal with the science/religion issue? In *Studies in Science Education* 44: 157-186.

Reiss, M.J. 2011. How should creationism and intelligent design be dealt with in the classroom? In *Journal of Philosophy of Education* 45: 399-415.

Reiss, M.J. In press. Beliefs and the value of evidence. In J.K. Gilbert & S.M. Stocklmayer (eds.). *Communication and engagement with science and technology: Issues and dilemmas.* London: Routledge.

Warnock, M. 2010. *Dishonest to God.* London: Continuum.

Watson, D.C.C. 1975. *The great Brain Robbery.* Worthing: Henry E. Walter.

Winslow, M.W., J.R. Staver & L.C. Scharmann. 2011. Evolution and personal religious belief: Christian university biology-related majors' search for reconciliation. In *Journal of Research in Science Teaching* 48: 1026-1049.

Current commentary: The arc of civil liberation[36]

Jeffrey C. Alexander

This book has offered differing perspectives on the concepts of science and religion, and the intertwining of their roles in modern societies. In the following I wish to take a closer look at three transformative political-symbolic experiences. They are Barack Obama, Tahrir Square, and the Occupy Wall Street movement. These cases serve as examples of how the sacred and profane is placed back into the idea of the secular, and show how they remain at the core of modern life. Our modern society is often perceived as rational, and traditional society as religious. Science, rationality and individual autonomy are given as characteristics of the ethos of modernity, juxtaposed as radically different from what they call traditional society. That is, we only believe what we see before our eyes – a kind of empirical sensibility – whereas traditional society was filled with mysteries. However, people still need broad, metaphysical beliefs and narratives, not necessarily in a super-natural world, but beliefs that are not proven empirically. No matter how sophisticated we are, no matter how much we say that we see ambiguity, we understand it, we can tolerate ambivalence, none the less I believe that certain core convictions are important in social organization. We still need sweeping social passions, symbols, the role of rituals and sentiments in contemporary and so-called secular society.

I will suggest that movements such as Obama's campaign for Presidency, the Egyptian demonstrations at Tahrir Square and the Occupy Wall Street movement embody an idea of the sacred. These movements should be seen not simply politically, as struggles for state power, but as symbolic upheavals in the spiritual hearts of their own nations and in other societies around the globe. Emotionally laden eruptions of utopian possibility, these performances wildly inspired their immediate participants, and projected "tableaux" beyond the scene, to tens of millions in the national and global citizen audiences who fused with the performances from outside.

These symbolic eruptions form a narrative arc, the sequential iteration of a utopian performance that, over recent decades, has become a deeply engrained culture structure in global civil society. This "global civil society social movement" (Khosrokhavar, in press) can be seen as a kind of recombinant social DNA. The utopian ideal of civil solidarity sits uneasily in a world of social inequality and

[36] This text is based on a Public Lecture given at Tulane University, 26 January 2012. The author thanks Diane Gramms for her suggestions.

individual restriction. Dissatisfaction with existing social arrangements is chronic. Civil society becomes restlessness. Episodes of liminality and demands for civil repair are the periodic result.

The utopian idea of a solidary community composed of autonomous yet mutually responsible citizens has been at the heart of Western modernity since the City States of the Renaissance. In the seventeenth and eighteenth century revolutions in England, America, and France, the civil imaginary was crystallized in democratic revolutions that made constitutionally regulated and self-governing communities of citizens the new rulers of their respective states. With the rise of industrial capitalism in the mid-nineteenth century, the program for political democracy came to be gradually displaced by the "social question," a focus on class inequality that pushed for socialism rather than democracy. Efforts to control the ravages of industrial capitalism and imperialism demanded the creation of enormous state bureaucracies. In the crush of these newly insistent interests, the civil society imperative was often pushed aside. Revolutionary strategy shifted from public mobilization to clandestine militancy, and violent political organizations became *de rigueur* on the left and right.

The most remarkable political development over the last four decades has been the withering of state-centrism and Jacobin ideals. Democracy has re-emerged as a radical idea and civil society as a revolutionary movement. In 1981, to the astonishment of liberal, radical, and conservative pundits alike, the "Solidarity" movement emerged in Poland. It was repressed the year after, but the decade that followed enshrined its idea of democratic civil society as a radical, revolution-inspiring goal. The blossoming of newly democratic Spain defied predictions that Franco's passing would trigger a bloody civil war. The "flower power" of the Philippine "People's Revolution" compelled Ferdinand Marcos to flee and the military to cede power to the million protesters in Manila's public square. Throughout the Southern Cone of Latin America, civilian governments pushed military juntas aside.

That first arc of global civil society movements culminated in the magical year of "1989," when one communist dictatorship after another fell before non-violent velvet revolutions. In June 1989, the communist state in China nearly met its match in Tiananmen Square. In 1990, pressure from global civil society compelled a peaceful transition to multicultural democracy in South Africa.

The 1980s created a new script for revolutionary social upheaval, one that left the utopia of socialism and the repertoire of violent militancy behind. Shifting from the proletariat to cross-class coalitions, from vanguard to mass participation, and from violence to nonviolence, the series of utopian uprisings made civil society seem radical. This story of liberation was constructed in familiar narrative patterns, as a movement of purity from danger, of light breaking through darkness, of enslaved peoples breaking their chains. But the characters who enacted this narrative had now changed. They took global politics in a new direction, back from 1917 and 1933 to 1789 and 1776. A new world revolution was being born (Sobral unpublished data).

In recent times, this narrative arc has been projected once again. It began with the national and worldwide effervescence of the Obama for President campaign in 2008, swept through North Africa and the Middle East in the spring and summer of 2011, and occupied Wall Street in the autumn of that year. The restless arc of civil utopia has not yet crested. Russia is on its trajectory as well.

"Obama"

One can explain the two-year campaign that Obama waged for the American Presidency as a struggle for political power, filled with strategy and money and ending with a resounding, if still relatively narrow, majority of votes. One can also understand these pre-Presidential Obama years as a utopian social movement. Obama's rise inspired tens of millions of Americans to hope and believe – in the unifying, egalitarian, and individually liberating possibilities of the civil sphere (Alexander 2010). The delirium of Obama's rallies marked liminal interruptions of public space, civil rituals that resounded with democratic effervescence. Obama's person became an iconic symbol radiating an aura of fundamental social change. His triumphal progress signaled inclusion over exclusion, hope over fear, civil solidarity over fragmentation, the victory of democratic justice over cynical resignation to the abuses of power. "O-BA-ma, O-BA-ma, O-BA-ma" was the call of a people's movement, of the civil sacred challenging the anti-civil profane, of purity winning out against danger, of the street beating out the establishment, of citizens organizing the defeat of money and institutions.

The difficulties encountered by Obama-in-Office should hardly be surprising. The utopian hopes his campaign stirred and embodied could never be satisfied by the mundane machinery of government. Indeed, Obama himself seemed the victim of his own utopian aspirations. The President seemed to believe that his political enemies would help him restore civil solidarity. Humiliated by political catastrophe, Republicans were prepared to do no such thing. Obama's dream of civil repair was defeated by brilliant Republican partisanship, which made a farce of his utopian aspirations.

Tahrir Square

It was only weeks after Republicans handed Obama his head on a platter – in the November 2010 Congressional elections – that the restless arc of civil social movement stretched to North Africa and the Middle East. Like the rise of Obama, the Arab Spring was totally unexpected. It was experienced as a volcanic eruption of almost foolhardy aspiration, and few believed it could be sustained. Yet, Tunisia's Jasmine revolution triggered a whole series of uprisings, the lava eventually flowing to Egypt, Libya, Yemen, Jordan, Morocco, Bahrain, and Syria. There had, in fact, been an intellectual revolution in the Arab world, an internal political-cultural development that, pushing back against Occidentalism, socialism, and violent Islamism, tentatively embraced the tenets of liberal if not secular democracy.

Yet it was in Tahrir Square, in Egypt, that this unexpected outpouring of radical democratic sentiment symbolically peaked. In this nation in the heart of the Arab world, the drama of democracy played out over 18 days. There were many hundreds of deaths and thousands of injuries, but the millions of protestors remained nonviolent. Tahrir Square became a microcosm of civil utopia (Alexander 2011). The January 25th movement didn't just protest and demand it, but performatively enacted it. The narratives of Tahrir projected by mainstream, alternative, and social media featured cross-class and cross-religious solidarities. Egypt was being born again, rising like a Phoenix from the suffering and humiliation of the Mubarak regime.

Whether this is a stillbirth has not yet been decided. The Egyptian army stepped aside during the eighteen days of revolt, promising to institutionalize the civil revolution. In the aftermath, they have become its greatest roadblock. Whatever the results, like "Obama" and the Polish Solidarity movements before it, Tahrir Square projected meanings in public far beyond the boundaries of the Egyptian nation state. The revolution's English Facebook page projected its narrative of civil revolution around the globe, receiving tens of thousands of wildly supportive posts. The Egyptian revolution captured the global imaginary; it became a neon-lit symbol of human courage, a flashing advertisement for the possibility of democratic social mobilization against the powers-that-be.

"The People Want the End of the Regime," "The People Want the End of Military Trials," "The People Want the Rule of Law" – these chants from Tahrir Square reverberated not only across the Middle Eastern and North Africa, but Europe and North America as well. After watching one Western government after another embrace the restrictive demands of fiscal austerity, massive demonstrations broke out in Madrid, in London, in Tel Aviv, and Madison. They were pushbacks of civil against market society, protests against the craven submission of democratic governments to failed corporate and financial elites. Among these demonstrators, explicit references to "Tahrir Square" frequently appeared. If Egypt had provided a live performance of civil power in the East, in the West it was now being replayed. The corporate-communist masters in China filtered out every Internet reference to the words "Egyptian revolution." Russia's rulers could not do the same, and they have lived to rue the day.

The Western iterations of Tahrir Square had distinctive icons and slogans. As for icons, Guy Fawkes made frequent appearances; the grinning white-faced anti-hero of the 1605 British "Gunpowder Plot" who had metamorphosed in the 2006 comic book *V for Vendetta*, which the Wachowski brothers turned into a commercially successful film (Sobral unpublished data). As for slogans, none approached "We are the 99%." As the arc of civil upheaval spread west and east from Cairo, the most potent poetic transliteration came from New York. As a retrospective in the *New York Times* (Kristoff 2011) put it: "The idea, according to some organizers, was to camp out for weeks or even months to replicate the kind, if not the scale, of protests that had erupted earlier in 2011 in places as varied as Egypt, Spain and Israel."

Occupy Wall Street

"Occupy Wall Street" was stunning and unexpected, a random electric spark that started a hot brush fire. The American left had been prostrate, the Tea Party in command and the Obama revolution seemingly in full retreat. Initially derided, the scruffy gathering of a few hundred protestors in Zuccotti Park soon became a catalyzing social event. Powerful ideological statements are metaphors, creating new relations among previously disparate social elements. Propelled by felicitous performances, ideological metaphors can make meaning in public in new and surprisingly consequential ways.

"Occupy Wall Street" was virtual, not literal, as brilliantly felicitous as any performance in a particular time and place can be. It symbolically thrust the critical, demanding, and egalitarian spirit of American democracy into the stultifying and musky chambers of elites. If its message and effect were symbolic, the performance itself was physically demanding. There were rain, tents, dirt, police attacks, and it lasted more than 60 days. Efforts to repress Occupy triggered immediate and effusive outpourings of public sympathy, the dirt and tents and the nonviolence and the human microphone gradually garnering grudging admiration.

By sticking it out, and publicly sticking it to the financial and corporate elite, Occupy embodied the ongoing struggle between civil and market society. Occupy had no real demands, but that was the point. Experts inside and outside the beltway had been churning out policy proposals for years. It was the performance of Occupy that was itself the achievement. Riveting citizen-audience attention well beyond the relatively narrow bandwidth of frustrated progressives, Occupy commanded the means of symbolic production – network and cable news, front pages of newspapers, and leading blog sites. It supplied its own facilities as well, live cam streams to cable TV, cell phone pictures leaping to websites around nation and world. Its gutsy, aggressive, yet determinedly civil performance of social justice earned Occupy a distinctive mystique, an aura of sacrality that provided protection against repressive moves from the state. The protests had the wind of public opinion in their backs. If Mubarak's army was afraid to intervene in Tahrir Square, how much more reluctant were the police forces of a relatively democratic state?

Zuccotti Park did not change policy, elect new representatives, or lower the unemployment rate.[37] What it did was create a vastly more energetic and critical form of civil power. One way to understand this upgrade is how it energized the left. Iterations of Occupy sprang up in more than 150 cities: Occupy Oakland, Occupy Los Angeles, Occupy Chicago, even Occupy New Haven and Yale. A coalition of 70 liberal organizations, the "American Dream Movement", formed to provide material and support.

But the impact of these liminal performances went beyond the audience on the left. It entered into the center of American collective consciousness. As a front-page article in the *New York Times* (Stelter 2011) put it: "The 99% Has Become an Ingrained part of the Cultural Lexicon." One percent and ninety-nine percent became magical numbers, culture structures that redistributed civil sacred and profane, morally re-weighting economic and political "realities." Long viewed as bungling but not venal, and certainly worth saving, the financial and corporate elite now became the vilified and polluted "One Percent". The masses of struggling Americans, formerly characterized as hapless objects – victims, shleps, and pretty much schmucks – were transformed into the purified "Ninety-Nine Percent," a collective agent demanding justice, a maligned hero finally fighting back.

> In a very short time, this movement has dramatically changed how we think about occupation. In early September, "occupy" signaled on-going military incursions. Now it signifies progressive political protest. It's no longer primarily about force of military power; instead it signifies standing up to

[37] See, e.g. the complaint by columnist Nicholas Kristof, New York Times, October 1, 2011.

> injustice, inequality and abuse of power. It's no longer about simply occupying a space; it's about transforming that space (Alim 2011).

One month after the occupation of Zuccotti Park, half of a national sample told pollsters that Occupy reflected the views of most Americans; and two-thirds of all those queried, including one-third of the Republicans, said the distribution of wealth needed to be made more equal (Kohut 2011). Three months later, a national survey reported that two-thirds of Americans now believed there were "strong conflicts" between the rich and poor, eclipsing divisions of race and immigration. Since 2009 there had been a 50 percent increase in this perception of class conflict, the largest increases reported among whites, middle-income, and independent voters, the latter presenting the most dramatic shift, from 23 to 68 percent (Stelter 2011).

As these post-Occupy effects began to be felt, right-wing Republican campaigning for their party's Presidential nomination began eviscerating Mitt Romney as a "vulture capitalist" for his work with Bain Capital. And the arc of utopian civil movement once again reached outside the United States. In October, the *New York Times* wrote that "demonstrations in emulation of Occupy Wall Street were held in Europe, Asia and the Americas, drawing crowds in the hundreds and the thousands." At the end of December, a radical leader of the Russian democracy movement evoked Occupy from his hospital bed. In a fiery speech projected on large screens outside on the Moscow streets, he called the assembled protesters "the 99 percent" and said Russia was led by a corrupt one percent of bureaucrats and oligarchs.

On January 22nd, investigating hospital privileges for the most affluent patients, the *Times* suggested that "in the age of Occupy Wall Street, catering to the rich can be trickier" (Bernstein 2012). It is because of the arc of civil liberation that we are in this new age.

Bibliography

Alexander, J.C. 2010. *The Performance of Politics: Obama's Victory and the Democratic Struggle for Power.* New York: Oxford University Press.

Alexander, J.C. 2011. *Performative Revolution in Egypt. An essay in Cultural Power.* New York: Bloomsbury.

Alim, S.H. 2011. What If We Occupied Language? The Opinionator (online). In *The New York Times*. December 21.

Bernstein, N. 2012. With Bulers and marble Baths, Hospitals Vie for the Affluent Ill. In *The New York Times*. January 22.

Khosrokhavar, F. (in press). *The New Arab Revolutions that Shook the World.* Boulder, CO: Paradigm.

Kohut, A. 2011. The "Haves" and the "Have-Nots". In *The New York Times*. October 18.

Kristoff, N.D. 2011. The Bankers and the Revolutionaries. In *The New York Times*. October 1.

Stelter, B. 2011. Camps Are Cleared, but "99 Percent" Still Occupies the Lexicon. In *The New York Times*. December 1.

Contributors

Jeffrey C. Alexander: Professor of Sociology at the Center for Cultural Sociology, Yale University. Most recent publication: *Performative Revolution in Egypt. An essay in Cultural Power*. New York: Bloomsbury, 2011.
E-mail: jeffrey.alexander@yale.edu

Ragnar Fjelland: Professor at the Centre for the Study of the Sciences and the Humanities (SVT), University of Bergen. Most recent publication: Niels Bohr: Fysiker og filosof. Introductory essay in *Niels Bohr: Atomteori og naturbeskrivelse*. Oslo: Bokklubbens kulturbibliotek, 2011.
E-mail: ragnar.fjelland@svt.uib.no

Cathrine Holst: Senior Researcher at ARENA – Centre for European Studies and Postdoctoral Fellow at the Department of Sociology and Human Geography, University of Oslo. Most recent publication: Gender democracy across Europe: reflections on explanatory factors and the input-outcome relationship. In Y. Galligan (ed.). *Deliberative Processes and Gender Democracy. Case Studies from Europe*. London: Routledge, 2012.
E-mail: cathrine.holst@arena.uio.no

Tone Lund-Olsen: Senior Executive Officer at the Centre for the Study of the Sciences and the Humanities (SVT), University of Bergen. Most recent publication: *Yearbook 2006, Bergen National Academy of the Arts*. Bergen: KHiB, 2006.
E-mail: tone.lund-olsen@uib.no

Ole Jacob Madsen: Guest Researcher at the Centre for the Study of the Sciences and the Humanities (SVT), University of Bergen. Most recent publication: Me at My Best: Therapeutic Ideals in Norwegian Women's Magazines. In *Communication, Culture & Critique* 5: 20-37, 2012.
E-mail: ole.madsen@svt.uib.no

Michael Reiss: Professor of Science Education at the Institute of Education, University of London. Most recent publication: How should creationism and intelligent design be dealt with in the classroom? In *Journal of Philosophy of Education* 45: 399-415, 2011.
E-mail: m.reiss@ioe.ac.uk

Kjetil Rommetveit: Associate Professor at the Centre for the Study of the Sciences and the Humanities (SVT), University of Bergen. Most recent publication: Genetic

enhancement, futures tense. In *Futures. The journal of policy, planning and futures studies*, 43 (1): 76-85, 2011.
E-mail: kjetil.rommetveit@svt.uib.no

Gunnar Skirbekk: Professor at the Centre for the Study of the Sciences and the Humanities (SVT), University of Bergen. Most recent publication: *Multiple Modernities. A Tale of Scandinavian Experiences.* Hong Kong: The Chinese University Press, 2011.
E-mail: gunnar.skirbekk@svt.uib.no

Barbara Herrnstein Smith: Professor Emerita at Duke University. Most recent publication: *Natural Reflections: Human Cognition at the Nexus of Science and Religion.* New Haven, CT: Yale University Press, 2010.
E-mail: bhsmith@duke.edu

Roger Strand: Professor at the Centre for the Study of the Sciences and the Humanities (SVT), University of Bergen. Most recent publication: Health Ideologies, Objectivism and the Common Good: On the Rights of Dissidents. In *Cambridge Quarterly of Healthcare Ethics*, 20 (4): 605-611, 2011.
Email: roger.strand@svt.uib.no

Nora Sørensen Vaage: PhD Candidate at the Centre for the Study of the Sciences and the Humanities (SVT), University of Bergen. Most recent publication: *Hybrids in art. Theoretical perspectives on art in the age of genetics.* Bergen Open Research Archive, 2011.
E-mail: nora.vaage@svt.uib.no

Simen Andersen Øyen: PhD Candidate at the Centre for the Study of the Sciences and the Humanities (SVT), University of Bergen. Most recent publication: *Humanioras fremtid.* Oslo: Cappelen Akademiske Forlag, 2011.
E-mail: simen.oyen@svt.uib.no

www.ingramcontent.com/pod-product-compliance
Lightning Source LLC
Chambersburg PA
CBHW061217240926
51846CB00006B/1255
9789086861965